An Unstable System

Ed Adams

a firstelement production

Ed Adams

First published in Great Britain in 2021 by firstelement
Copyright © 2021 Ed Adams
Directed by thesixtwenty

10 9 8 7 6 5 4 3 2 1

A CIP catalogue record for this book is available from the British
Library.

ISBN 13: 978-1-913818-16-6

eBook ISBN: 978-1-913818-17-3

Printed and bound in Great Britain by Ingram Spark

rashbre
an imprint of firstelement.co.uk
rashbre@mac.com

ed-adams.net

For Personas, Presences and
those that live in the Question.

Ed Adams

Thanks

A big thank you for the tolerance and bemused support from all of those around me. To those who know when it is time to say, " step away from the keyboard!" and to those who don't.

To Julie for understanding that only comes with really knowing me.

To thesixtwenty.co.uk for direction.

To the NaNoWriMo gang for the continued inspiration and encouragement.

To Topsham, for being lovely.

To John, for his uncompromising readthroughs.

To Steve, for his boundless encouragement.

To Elizabeth, Georgina, Roger, John, Caroline, Richard and other cover reviewers.

To the edge-walkers. They know who they are.

And, of course, thanks to the extensive support via the random scribbles of rashbre via
http://rashbre2.blogspot.com
and its cast of amazing and varied readers whether human, twittery, smoky, cool kats, photographic, dramatic, musical, anagrammed, globalized or simply maxed-out.

Not forgetting the cast of characters involved in producing this; they all have virtual lives of their own.

And of course, to you, dear reader, for at least 'giving it a go'.

Books by Ed Adams include:

Triangle Trilogy		About
1	The Triangle	Dirty money? Here's how to clean it
2	The Square	Weapons of Mass Destruction – don't let them get on your nerves
3	The Circle	The desert is no place to get lost
4	The Ox Stunner	The Triangle Trilogy – thick enough to stun an ox
		(all feature Jake, Bigsy, Clare, Chuck Manners)
Archangel Collection		
1	Archangel	Sometimes I am necessary
2	Raven	An eye that sees all between darkness and light
3	Card Game	Throwing oil on a troubled market
4	Magazine Clip	the above three in one heavy book.
5	Play On, Christina Nott	Christina Nott, on Tour for the FSB
6	Corrupt	Trouble at the House
7	An Unstable System	
		(all feature Jake, Bigsy, Clare, Chuck Manners)
Stand-Alone Novels		
1	Coin	Get rich quick with Cybercash – just don't tell GCHQ
2	Pulse	Want more? Just stay away from the edge
3	Edge	Power can't be left to trust
4	Now the Science	the above three in one heavy book.
Blade's Edge Trilogy		
1	Edge	World end climate collapse and sham discovered during magnetite mining from Jupiter's moon Ganymede.
2	Edge Blue	A human outcome, after a doomsday reckoning, unless…
3	Edge Red	An artificially intelligent outcome, unless…
4	Edge of Forever	Edge Trilogy

About Ed Adams Novels:

Triangle Trilogy		About
1	Triangle	Money laundering within an international setting.
2	Square	A viral nerve agent being shipped by terrorists and WMDs
3	Circle	In the Arizona deserts, with the Navajo; about missiles stolen from storage.
4	Ox Stunner	the above three in one heavy book.
		(all feature Jake, Bigsy, Clare, Chuck Manners)
Archangel Collection		
1	Archangel	Biographical adventures of Russian trained Archangel, who, as Christina Nott, threads her way through other Triangle novels.
2	Raven	Big business gone bad and being a freemason won't absolve you
3	Card Game	Raven Pt 2 – Russian oligarchs attempt to take control
4	Magazine Clip	the above three in one heavy book.
5	Play On, Christina Nott	Christina Nott, on Tour for the FSB
6	Corrupt	Parliamentary corruption
7	An Unstable System	
		(all feature Jake, Bigsy, Clare, Chuck Manners)
Now the Science Collection		
1	Coin	cyber cash manipulation by the Russian state.
2	Pulse	Sci-Fi dystopian blood management with nano-bots
3	Edge	World end climate collapse and sham discovered during magnetite mining from Jupiter's moon Ganymede
4	Now the Science	the above three in one heavy book.
Blade's Edge Trilogy		
1	Edge	World end climate collapse and sham discovered during magnetite mining from Jupiter's moon Ganymede.
2	Edge Blue	Endgame, for Earth – unless?
3	Edge Red	Museum Earth – unless?
4	Edge of Forever	Edge Trilogy

Ed Adams Novels: Links

Triangle Trilogy		Link:	Read?
1	Triangle	https://amzn.to/3c6zRMu	
2	Square	https://amzn.to/3sEiKYx	
3	Circle	https://amzn.to/3qLavYZ	
4	Ox Stunner	https://amzn.to/3sHxlgh	
		(all feature Jake, Bigsy, Clare, Chuck Manners)	
Archangel Collection			
1	Archangel	https://amzn.to/2Y9nB5K	
2	Raven	https://amzn.to/2MiGVe6	
3	Raven's Card	https://amzn.to/2Y8HLgs	
4	Magazine Clip	https://amzn.to/3pbBJYn	
5	Play On, Christina Nott	https://amzn.to/2MbkuHl	
6	Corrupt	https://amzn.to/2M0HnOw	
7	An Unstable System		
		(all feature Jake, Bigsy, Clare, Chuck Manners)	
Now the Science Collection			
1	Coin	https://amzn.to/3o82wmS	
2	Pulse	https://amzn.to/3qQlBvL	
3	Edge	https://amzn.to/2KDmYOW	
4	Now the Science	https://amzn.to/3iG5Nc2	
Edge of forever Trilogy			
1	Edge	https://amzn.to/2KDmYOW	
2	Edge Blue	https://amzn.to/2Kyq9au	
3	Edge Red	https://amzn.to/2KzJwjz	
4	Edge of Forever	https://amzn.to/3c57Ghj	

TABLE OF CONTENTS

Ed Adams

PART ONE

Beauty and Terror

"Let everything happen to you
Beauty and terror
Just keep going
No feeling is final."

Rainer Maria Rilke

Author's Note

This book tells the story of Matt Nicholson, who was the brains behind the original Coin cyber coin manufacturing device. After he gained certain wealth, he decided he would turn his full attention towards innovation and it is the consequences we see reported here.

I've cleaned up his wording somewhat and occasionally skipped a graphic moment, but most of what is in the following pages is as Matt intends it to be. Matt thinks of it as notes on part of a life.

Oh yes, and don't try any of this at home.

Prologue

Described in Coin:

Matt Nicholson, Tyler Sloan and Kyle Adler had shared a flat as students. Matt devised a cyber coin mining system, and the others helped with the invention, with it ultimately receiving the attention of Grace Fielding at GCHQ and Amanda Miller and Jim Cavendish at MI6.

The coin-mining device was exceptional and had been used to expose currency manipulations driven by Russia and with interference from the United States.

Erica Sweeney, an ex-girlfriend of Tyler, worked in a commercial bank and had noticed the signs of large-scale fraud, being operated by her boss Victor Boyd. Boyd disappeared, as did Marcus Barton, the analyst boss of Tyler Sloan. Rosie Marr, Tyler and their associates at GCHQ and SI6 needed to follow the trail and ultimately to handle the damage from the Coin machine. The aftermath took its toll on Matt, whose story picks up after these events, and when he has come by a large sum of residual money from the Coin system.

Ed Adams

What's He Building?

What's he building in there?
What the hell is he building in there?
He has subscriptions to those magazines
He never waves when he goes by

He's hiding something from the rest of us
He's all to himself
I think I know why
He took down the tire swing from the Peppertree
He has no children of his own, you see

He has no dog
And he has no friends
And his lawn is dying
And what about all those packages he sends?

Tom Waits

Hook

There are a few things I didn't ever tell the others about the invention of the cyber-mining device.

The most obvious one is the way that I was boosting my thought processes. Those that know me will understand. Writers and other artists sometimes use substances to boost their creativity.

Native American Indians said that peyote took them to heaven, but white missionaries would say, with equal assurance, that it offered them only a glimpse of hell.

Jean-Paul Sartre tried mescaline, and according to his companion Simone de Beauvoir, had a very bad trip: 'The objects he looked at changed their appearance in the most horrifying manner: umbrellas had become vultures, shoes turned into skeletons, and faces acquired monstrous characteristics...'

By the 1950s, Aldous Huxley was writing Doors of Perception and Heaven and Hell under the influence of mescaline, the synthesised version of peyote. Then he moved to LSD, which became available to adventurous writers, intellectuals, and therapists.

William Burroughs and the Beat writers of the 1950s and 60s reconfigured the psychedelic landscape by moving hallucinogens out of the drawing room and into the streets, pursuing their organic roots in the third world.

Burroughs wrote portions of Naked Lunch under the influence of yage, or ayahuasca, the DMT-containing hallucinogenic brew concocted in South America: 'New races as yet unconceived and unborn, combinations not yet realised pass through your body. Migrations, incredible journeys through deserts and jungles and mountains... The Composite City where all human potentials are spread out in a vast silent market.'

Allen Ginsberg, the beat poet, took peyote in Mexico and yage in South America. His poem Junky describes Burroughs's peyote experiences, and portions of Ginsberg's epic poem Howl were also written under the influence of peyote.

And we shouldn't forget Earnest Hemingway, who first coined the phrase 'Write Drunk, Edit Sober.'

My approach whilst trying out new ideas had some similarities, although I used electronic instead of chemical stimulation. I'd seen several brain booster devices on eBay and in Wired magazine and decided 'How difficult can it be?' to make one.

The technical term is transcranial direct current stimulation (or tDCS), and it involves hooking up electrodes to the skull and then turning on a small electric current, typically powered by a 9-volt battery.

There's a small community citing this as its inspiration. Some studies that have found tentative promise for tDCS to enhance memory, alertness, and the ability to learn new tasks, and to decrease symptoms of anxiety and depression. Anecdotally, users report that tDCS also helps them ease into a flow state (i.e., being "in the zone"), where they can get many tasks done without distraction. Others will do this by listening to Mozart.

Of course, I'd read the literature. The jury was divided. Some said it would boost thinking, others said it would create limits. There were further warnings about the voltages. It wasn't like overclocking a cpu. The desired maximum voltage seemed to be 9 volts or the equivalent output of a single PP3 battery. That's oblong battery you find in the old-fashioned smoke detectors. Then it would be at around 1 or 2-milli-Amperes and to run it for about 10 minutes.

I priced up some components and ordered them from a couple of electronic specialists. The electro-pads were hardest to get, and suppliers wanted to prove they were medically certified, which added cost. I improvised instead with some Medium Wave ribbon antenna and a few Band-Aids. The entire system, including power transistors and some Veroboard, cost me less than £20.

I'll be honest. At the time I didn't want the others in the flat to see me wired up, so I hid everything around the back of a chest of drawers, which was conveniently in the middle of the floor in my room. And I had the whirring clicking machinery of the cyber coin miner on the other cupboard which made a great distraction.

Now I could fulfil my dream to play Tom Waits and attempt to jack my brainpower at the same time. Maybe listening to "What's he building in there?"

Rats with backpacks

I had been playing around with the tDCS for a while when I found an old black and white documentary about brain stimulation in cats.

It made me think of other obvious applications, and I looked around for the literature on rats. Sure enough, there was plenty of experimental data around, but I felt somehow cheated as I read it.

It wasn't really doing anything particularly clever, and I felt that the whole invasive procedure wasn't kind to the animals. It used micro-electrodes implanted into their brains and relied on stimulating the reward centre of the rat.

There were three electrodes implanted; two in the ventral posterolateral nucleus of the thalamus which conveys facial sensory information from the left and right whiskers, and a third in the medial forebrain bundle which is involved in the reward process of the rat.

The third electrode was used to give a rewarding

electrical stimulus to the brain when the rat makes the correct move to the left or right. During training, the operator stimulates the left or right electrode of the rat making it "feel" a touch to the corresponding set of whiskers, as though it had met an obstacle. If the rat then makes the correct response, the operator rewards the rat by stimulating the third electrode. It's appalling, and yet another example of how the human species instrumentalises other species.

I felt that this was too much like using a traumatic invisible electric fence to control the rat, with left or right whiskers stimulated and an appropriate reward signal. Not cool.

Then in 2002, a team of scientists at the State University of New York remotely controlled rats from a laptop up to 500 metres away. They could instruct the rats to turn left or right, climb trees and ladders, navigate piles of rubble, and jump from different heights. They could even be commanded into brightly lit areas, which rats usually avoid. It was suggested that the rats could carry cameras to people trapped in disaster zones.

Things progressed slowly and in 2013, researchers reported the development of a radio-telemetry system to control free-roaming rats with a range of 200 metres. The backpack worn by the rat includes the mainboard and an FM transmitter-receiver, which can generate biphasic micro-current pulses. All components in the system are commercially available and fabricated from surface mount devices to reduce the size and weight (down to 10 grams)

Concerns have been raised about the ethics of such creepy studies.

An expert in animal welfare law at Rutgers University School of Law, said 'The animal is no longer functioning as an animal, because the rat is operating under someone's control.' And the issue goes beyond whether the stimulations are compelling or rewarding the rat to act. "There's got to be a level of discomfort in implanting these electrodes," he says, which may be difficult to justify. We are also faced with the animal's native intelligence or instinct which can stop it from performing some directives but with enough stimulation, this hesitation can sometimes be overcome, but occasionally cannot. Full-on conditioning.

Then I looked for some non-invasive methods, which were terribly cumbersome. Researchers at Harvard University created a brain-to-brain interface (BBI) between a human and a Sprague-Daley rat.

Simply by thinking the appropriate thought, the BBI allows the human to control the rat's tail. The human wears an EEG-based brain-to-computer interface (BCI), while the anaesthetised rat is equipped with a focused ultrasound (FUS) computer-to-brain interface (CBI). FUS is a technology that allows the researchers to excite a specific region of neurons in the rat's brain using an ultrasound signal (350 kHz ultrasound frequency, tone burst duration of 0.5 ms, pulse repetition frequency of 1 kHz, given for 300 ms duration).

The main advantage of FUS is that, unlike most brain-stimulation techniques, it is non-invasive. Whenever the human looks at a specific pattern (strobe light flicker) on a computer screen, the BCI communicates a command to the rat's CBI, which causes ultrasound to be beamed into the region of the rat's motor cortex responsible for tail movement. The researchers report that the human BCI has an accuracy of 94%, and that it generally takes around 1.5 s from the human looking at the screen to the movement of the rat's tail.

This was utterly clunky as an approach and seemed to illustrate we were a long way from anything faintly practical.

Another system that non-invasively controls rats uses ultrasonic, epidermal, and LED photic stimulators on the back. The system receives commands to deliver specified electrical stimulations to the hearing, pain and visual senses of the rat respectively. The three stimuli work in groups for the rat navigation.

So, it looks like 'rats with backpacks', is about as far as that line of enquiry has reached.

Heather

I first met Heather when I visited one of the labs as part of my ongoing interest in cybersystems. She gave the appearance of a youthful student, with her shoulder-length brown hair, black jeans, and a blue long-sleeved tee-shirt. Heather showed me around the facility in Cork, Ireland. She was running the animal inventory. I could see she really cared for the critters in her charge, and most of even the smallest insects had pet names.

In her research lab, the animals were predominantly controlled using radio signals. The electrodes signalled a direction or action desired by the human operator and then stimulated the animal's reward centres if the animal complied. Tiny backpacks had been devised, even for the smallest creature.

These animals are sometimes called bio-robots or robo-animals. They can be considered as cyborgs because they combine electronic devices with an organic life form. With the surgery required, and the moral and ethical issues, there has been criticism aimed at the use of remote control animals, especially regarding animal welfare and animal rights. Heather's lab included beetles, cockroaches, rats, dogfish sharks, mice, and pigeons.

The remote-control animals were designated and used as working animals for search and rescue operations or other specialised uses (which euphemistically referred to military uses).

The smallest successful cyborgs that the lab had produced were cockroaches, with a complete backpack of bright red circuit boards and tiny battery. These small creatures were comparatively docile and could be made to operate in the most alien of terrain. Inevitably they had been equipped with cameras and listening devices to make an ultimate and semi-autonomous bug.

I guess these animals were the luck ones in the wider lab. Other animals (mainly mice) were being used in tests by other departments, predominantly for botox-related products. A far cry from the days of the smoking beagles.

We'd gone out for our first date, along to Charlie's bar, close to the river. We'd both ordered pizza and were quietly chatting.

"Whereabouts in Scotland are you from originally?" I asked.

"Peterculter, near to Aberdeen. My father worked the North Sea rigs - although now he has a desk job. My mother is a doctor at an Aberdeen hospital. Oh yes, and my brother David left Scotland to pursue a career in banking down south."

"And how did you wind up here in Ireland?" I asked.

"My interest in physics and engineering led me into robotics and so I left Aberdeen for Strathclyde University near Glasgow - I got a First and then followed my brother south to London to study for an MSc at University College, part of the University of London."

"That's when I met Michael, an Irish boy, and after we'd both graduated, we came back to his hometown of Cork, where Michael got a job at the burgeoning Cork Science and Innovation Park in Curraheen. He knew a few others there and got in easily enough. What's your story?"

I considered how much to say about my time in London with the cyber currency-mining, GCHQ, the Russians and so on. It all seemed somewhat unbelievable.

I really liked Heather. Her intelligence and empathy really shone through. We'd stuck up an immediate friendship and almost before I know it, I was making plans to stay over in Cork.

Heather continued, "Yes, the proximity of the robot labs enticed me as well as Michael, and I got a job but not at Michael's firm. That would be too weird. I got a job here with a research component, even if it was secondary to the critter wrangling. I stayed here, even after the failure of my relationship with Michael. It all got messy when he fell for another woman at the lab."

Ouch.

Over the next few months Heather and I made some

plans and although I never mentioned my supply of money from the cyber coin mining, now some of it came in useful because we could get an apartment together.

I also rather dumbly hadn't realised that Cork was one of the biggest MedTech centres in Europe and had scientists from Johnson & Johnson, Pfizer, Lilly, Abbott, Stryker and many others. Heather got me an interview with a Chief Scientist at her facility, but I really used it as a springboard to practice my interview technique and to find somewhere with the best R&D profile.

That's how I came to be working for the well-funded Brant Industries. They were setting up a new subsidiary in Cork and had additional funding from Raven Corp. Their spending on high technology equipment was insane! They even dug out to install a 6.3 kilometre circumference 900 Giga-electron-volt Super Proton Synchrotron.

The only thing was, they liked me at Brant and now wanted me to go to one of their other facilities, where I could extend my Research brief. I'd have to tell Heather.

American visitor

As part of my recruitment to the other, bigger R&D facility, they invited me to a dinner with Bob Ranzino, a Senior Vice President of Brant, based out of Greenwich, Connecticut. Bob was a typical head office type. He had the expensive sweater and shirt (mercifully un-branded and un-monogrammed). His suntan was clearly from several vacations and he could talk lakeshores and yachts with aplomb.

We were sitting in the restaurant at Greene's in Cork, a fancy restaurant by Cork's standards. Bob had picked the seared hake and I went for the feather blade of Angus. Bob also selected the wines, and the sommelier was bringing it by the glass, so that we'd have something relevant to our main dish. Mine was Montepulciano, his was Viognier although where I come from, we call it Savvy B.

Bob had also brought an HR person along - Jasmine Summers. An attractive, auburn-haired willowy woman, wearing a blue, textured trench coat dress with a gold belt. She was turning heads in the restaurant but remained remarkably quiet, with the same meal selection as me.

I wondered why Bob had picked a large horseshoe curved banquette for our seating and then why Jasmine and Bob seemed to be squashed up to one side, facing across to me on the other. It really felt like an interview.

"So how are you finding the life at Brant?" he began.

"Oh, it's good to work somewhere that is prepared to invest in full-scale Research and Development," I replied.

"Our HR Department had a look at you, from public records, of course," he said, with twinkle in his eye, "That's right, isn't it, Jasmine?"

Jasmine nodded, "Yes, you were remarkably difficult to track down on the internet!"

Bob continued, "Nonetheless, I can see you are quite the inventor!"

I nodded and wondered how much they had found out about me.

"That student project, to make a robot room cleaner, it was pretty good," he said.

I realised that the HR Department had found the story about the robotically adapted Bex-Bissell carpet sweeper from my student days. It didn't sound as if they had found out anything about the cyber coin - Amanda Miller and Grace Fielding must have done a good job to hide everything.

"Oh yes, I took one of those manual carpet sweepers, added a couple of beefy motors, servos, sensors and a small camera rig, and the rest, as we know is Roomba's history!"

"I seem to remember that the device I'd built was surprisingly tough, but the oblong nature of it made it difficult to steer. On the other hand it had enough power to push furniture around. I guess it would have been better in Robot Wars with a chain saw on the front."

Bob smiled, but I don't think he knew what Bex-Bissell, nor Roomba nor Robot Wars was about.

"I guess it goes to show my early involvement in robotics," I said, "It is a lot more interesting now that we have Artificial Intelligence as well. And I guess we could consider humans to be part-cyborg too."

"How so?" asked Bob, plainly intrigued.

"Well, humans augment their own brainpower with other digital devices. Things like smartphones or laptops, increasingly with devices like Alexa. The challenge is that humans are I/O bound."

"Input/Output bound? What do you mean?" asked Bob.

"Well, I might have good input systems, like high-definition, high bandwidth vision through my eyes, but my output systems are somewhat slower. Like my typing or speaking speed. Maybe 120 words per minute. That's not very fast, is it?"

I could see Jasmine looking confused.

Bob smiled, "Yes, you are right, we'd need to speed up the output to make things really useful, although don't you think there would be a buffering problem?"

"Externally, yes, if I could really output quickly, then the ability for the information to be absorbed by another might be compromised. Like, in PowerPoint, when we say 'drinking from the firehose!'"

Bob answered, "Yes, I can see that glazed look sometimes when I do company presentations; I'm never sure whether its boredom or insouciance."

"Or people expect to be interrupt-driven nowadays. They can't concentrate on a single task for more than a few minutes without needing click-gratification. Maybe you could gamify your pitches? Add something that they need to count, where points mean prizes?"

"Interesting, I'll have to bear that in mind for my next All-Company presentation!"

Jasmine wrote something in her small black notebook.

"So let me tell you what we've been thinking about over at Brant's main R&D facility. It's where we study HCI - Human-Computer Interfaces."

"What about HCCI?" I asked, "The next level - where a human uses a machine protocol to address another human in real time? I suspect it will be the basis of cyborg systems in the future."

"That's good. It's also what we are thinking about now. Ways to improve the human through augmented systems. It could be AI, or AR, or AO."

"Artificial Intelligence, Augmented Reality and - er - I'm guessing this last one Augmented Organics?" I asked.

Bob looked impressed that I'd figured it out so quickly.

"It's the free-will element," I said, "Think about a simple organism, like a cockroach, with a computer backpack. You can stimulate it to move left and right, reward it even, but put it in front of some flames and tell it to march on and what will it do? It'll turn around to preserve itself. No matter how many stimulation-kicks it gets, it will still want to run away from the flames."

Bob looked impressed again, "Have you studied this?" he asked.

"No," I replied, "But my girlfriend works in a lab. They have critters." I looked over to Jasmine to make sure she was writing this down.

"Oh, sorry to have been talking about this whilst we are trying to eat. These steaks look so well finished too."

Bob paused, "I think we'll want to make an offer to re-assign you. You'll get the whizz-bang facilities and a rather extensive budget, but I guess you'll want to talk it over with Heather?'

I noticed immediately that'd mentioned Heather by name, although I had not. Either Jasmine was suddenly good at her job or Bob had found out by some other means.

"Would it be an assignment or a transfer?" I asked.

"You should consider it to be a long-term transfer, to one of the best facilities of its kind in the world," answered Bob.

"Where?" I asked.

"Geneva, Switzerland," answered Bob, "We'd find you accommodation and visas etc. Do you speak French?"

Emotional Intelligence

I needed to explain to Heather about my upcoming transfer to Geneva. The fundamental problem was that Geneva, Switzerland required special work visas.

Heather wouldn't be eligible and could only start there as a visitor - a guest of me. Despite a decent brain, I suspect I have a low EQ Emotional Quotient - My emotional intelligence isn't quite up to scratch. I know I should be able to understand, use, and manage my emotions in positive ways to relieve stress, communicate effectively, empathise with others, overcome challenges and defuse conflict.

I know all the theory - If I had a great emotional intelligence it could help me build stronger relationships, succeed at work, and achieve my career and personal goals. It could also help me connect with my feelings, turn intention into action, and make informed decisions about what matters most to you.

You know that noise that people make when they push a needle across a vinyl record. Sort of zzzrp. Imagine that sound now.

We were sitting in the Cork flat, drinking coffee, either side of a small table in the sitting room.

"So why did they offer this role to you?" asked Heather.

'Self-management,' I thought, *'I should be able to control impulsive feelings and behaviours, manage my emotions in healthy ways, take initiative, follow through on commitments, and adapt to changing circumstances.'*

"They must have thought I was good. They offered me the job over a steak in Greene's," I answered, "That head honcho named Bob Ranzino had come over from America specially. It sounds great; R and D with all the latest technology, I'd be able to follow my research dreams."

"Did they ask you about me?" asked Heather.

'Self-awareness,' I thought, *'I should recognise my own emotions and how they affect my thoughts and behaviour. I know my strengths and weaknesses, and have self-confidence.'*

"No, they didn't, but I mentioned you to them," I answered, "I said thought you'd see my decision as a no-brainer,"

"Did you have a plan about all of this, then?" asked Heather. She was looking annoyed. I could see her hand shaking as she poured another cup of coffee.

'Social awareness,' I thought, *'I have empathy. I can*

understand the emotions, needs, and concerns of other people, pick up on emotional cues, feel comfortable socially, and recognise the power dynamics in a group or organisation.'

"Not really, but I thought that a change of location could be a great way to clear the air!"

"Clear the air? What does that mean? Do you even still care about me?" asked Heather. I worried that the coffee was making her irritable.

'Relationship management,' I thought, *'I know how to develop and maintain good relationships, communicate clearly, inspire and influence others, work well in a team, and manage conflict.'*

"Of course I do, but I can see you are not too happy about this great opportunity for me!"

Heather replied, "I can see this is all about you, but I'd hoped there was enough room in our relationship for it to be about me, too!"

As we know, it's not the smartest people who are the most successful or the most fulfilled in life. We probably know people who are academically brilliant and yet are socially inept and unsuccessful at work or in their personal relationships. Heather was showing all the signs - either that or it was the Java.

Intellectual ability and intelligence quotient (IQ) aren't enough on their own to achieve success in life. Yes, IQ can help get into college, but it's the EQ that will help manage the stress and emotions when facing final exams.

IQ and EQ exist in tandem and are most effective when they build off one another.

If only I could have remembered that when I was talking to Heather. So what happened next?

"Well, if that's how you feel about it, then I'm going out for a while until you've cooled down!" I said.

I texted Danny, my friend from work. We agreed to meet in the Oliver Plunkett.

"Man, what are you doing? Heather's a lovely girl!" said Danny, "What d'ya want to go to Switzerland for, anyway?"

"I think I need to push myself," I said, "They saw me as a significant job candidate, after all. You could even say I was head-hunted over steak at Greene's"

"But won't the emotion of all this get you down and stop you from operating effectively? I mean, if you and Heather split up? And it is a new place, and a new job, even a different language?" asked Danny.

"No, the Brant labs have a house language of English. It's because they employ so many nationalities there," I thought I had neatly sidestepped Danny's other questions.

He was looking at me strangely aghast. I didn't know what that meant.

"I can manage my emotions," I said after a short pause,

"After all, if I wasn't able to manage emotions, I probably couldn't manage stress either. That could lead to serious health problems. Uncontrolled stress would raise my blood pressure, suppress my immune system, increase the risk of heart attacks and strokes, contribute to infertility, and speed up the ageing process. The first step to improving emotional intelligence is to learn how to manage stress."

"I can see that's an area you'll be working on, to be sure," said Danny, "You don't want it to start messing with your head!"

I answered, "My mental health is in good shape. Uncontrolled emotions and stress would affect my mental health and make me vulnerable to anxiety and depression. If I'm unable to understand, get comfortable with, or manage my emotions, I'll also struggle to form strong relationships. This could leave me feeling lonely and isolated. No, I'll keep my mental health in good shape.

Danny said, "You say that, but over here, in Cork, you've also got friends around you. People like me whose shoulder you can cry on when things get tough. And the good lady, Heather herself!"

I answered, "No, I think that by understanding my emotions and how to control them, I'm better able to express how I feel and understand how others are feeling. This should allow me to communicate more effectively and forge stronger relationships, both at work and in my personal life. That includes in any new life I decide to create, such as in Switzerland."

I was pleased with my social intelligence. Being in tune with my emotions served a strong social purpose, connected me to other people and the world around. Social intelligence enabled me to recognise friend from foe, measure another person's interest in me, reduce my stress, balance my nervous system through social communication, and to make me feel loved and happy.

"I'm all over this," I thought.

Geneva and plum flan

I arranged to keep the apartment in Cork, for another six months, so that Heather could work out what she wanted to do. My Emotional Intelligence had failed, and Heather had a sense of humour failure about my entire adventure. Her parting words to me were, "Va te faire foutre!"

I checked the flights to Geneva. The Irish national budget airline wasn't very helpful for this flight. It could get me to London and then I'd have to pay full fare to fly on to Geneva, after a long layover.

It was better to fly with the Dutch airline, which had a shorter stopover in Amsterdam and was one-fifth the cost of the Irish/British flight.

Flying into Geneva, there's a magnificent view of the lake, and I could even see the fountain from the air. What is also striking, is the number of houses with pools in what should, by rights, be a colder climate.

Then Geneva airport, which has changed since the last

time I visited. Most airports are in a continuous state of construction work and Geneva had been no exception, with a huge makeover, although I could still recognise some routes and walkways, especially once I got to the landside of the airport.

Geneva is also one of those airports with arrivals Duty-Free (like Heathrow) although I hurried through the entire system even in the knowledge of Swiss Prices being even higher than those in London.

I was a few days ahead of my official start at the labs and had already rented an apartment in central Geneva, although the cost, at CHF 3,100 was mad, even compared with London prices. That was for 120m2 and two bedrooms. Fortunately, I had my cyber coin money as well as the new salary which I'd be given by Brant. They'd given me a CHF 10,000 relocation expenses 'float' to get me started, but I soon realised that it should be called a 'melt' in Switzerland. As I passed by McDonalds, I noticed a McMeal cost CHF 14.50, which I reckoned to be around $17.00. That's quite a lot for a burger, fries, and a coke.

I could feel that little stab into my brain, like when I'm with buddies on the border of one too many drinks. It was saying, "Careful! Careful!"

I queued for a taxi outside the airport and asked for my new address in Geneva. Rue de la Confédération. I hoped the driver knew the way (he did) and so I didn't need to point awkwardly at the map which I'd printed out and had in my backpack.

I'd called the perfectly delightful Aude who was the Courtière Locations Résidentielles and said she would meet me at the Apartment and show me the way in and how the security system worked. The taxi driver had put my two heavy bags into the back of the car, and I was now being whisked in a Mercedes taxi across town to my new home.

Around twenty minutes later, my taxi pulled up outside an apartment block. I knew we were in the centre of the city, on the south side of the Rhone and I looked around for Aude.

I decided to call her, as the taxi driver unloaded my bags from the back of the car. In good English he then asked for the fare, which was CHF160. I rapidly calculated that it was around US$180, from the airport to the centre. £131 for 25 minutes or 6 kilometres. My float was melting fast.

Aude appeared, smiling, from around a corner.

"Hello Mr Nicholson, welcome to Geneva; Welcome to Switzerland. My name is Aude Darmshausen," She spoke in perfect English, with a hint of a posh Home Counties accent.

"Hello," I said, "It's great to finally meet you! Forgive me for asking, but are you Swiss, only you have such a very good English accent?"

"Yes, I'm Swiss, but I get asked that a lot by Brits. My family lived in West London when I was growing up; I spent 6 years there, in Chiswick. Maybe you know it?"

"Yes, I know Chiswick very well, I used to live in Kensington and we'd sometimes go out that way to meet up with some friends - usually along the river. Strand on the Green."

"It's a small world, " replied Aude, "Let's get you inside and to the apartment. It is quiet today because it's a public holiday. 'Jeûne genevois' - translated to 'Genevan fast' in English - a religious holiday that is now more traditionally accompanied by eating a plum pie.

"This apartment is in a lovely location for all of Geneva. You can walk to the centre in less than ten minutes, or to the fountain and the Gardens in maybe five."

At the main door she paused and held an entry card to an impressive card reader on the wall.

"The security was updated recently," she said.

The door buzzed, and she pressed it open.

Then into the elevator and we exited in a corridor which was tiled, both on the floor and the walls.

"Interesting decor!" I said.

"Remember, we get snow here," answered Aude, "The corridors take the traffic of snow and ice. The brown and orange colours are everywhere. More Germanic than French, though."

She led me to the Apartment. Another hotel room style key entry system.

"I'm going to leave you with three keys. I suggest you carry one, put one safe somewhere outside the apartment, and keep the other one in the kitchen drawer. And look, you have an entry phone system with a camera. Press the big green button to let people in."

"Thanks, that's good advice," I said.

"You can get extra keys made, if you so wish, just don't tell me about them," said Aude, "Insurance, you understand."

I looked around the Apartment. They had furnished it in a modern, cool style.

"Wow. I wasn't expecting this. It's so modern inside!"

"I hope you like the interior decor more than that outside?" asked Aude, "I had to specify most of this myself. You were renting an unfurnished apartment and needed it turned into a turnkey. And the reason it looks so modern is that this entire block was refurbished around two years ago. New plumbing, electrics, everything."

"It's great, fantastic, quite stylish. I wish I knew the person who lived here," I said, "That kitchen: it's amazing!"

Aude smiled, "Thank you, I've got everything down to the cutlery and utensils as well. Some smaller items are from Bongénie Grieder and the rest from Ikea. I thought you'd like a few fancy items from Bongénie mixed in with IKEA's utility."

"This is fantastic," I said, "What is that white thing on the kitchen counter?"

"Oh, a Presse-agrumes électrique," answered Aude, "It's by Alessi, like the bottle opener and the Bouilloire électrique - electric kettle - to make proper British tea!"

"How did you get all of this?" I asked.

"Oh, me and my boyfriend Marc spent a weekend 'shopping' for the smaller items. I ordered the main things online. Brant had said there was a practical budget but they knew they would have to adjust it for the Swiss cost of living. I think all of this came to less than CHF 12,000."

"I can't thank you enough, Aude,"

"I've arranged for a weekly cleaner to come along too, it's from a firm called Batmaid. They are well-known here in Geneva and have many reliable cleaners. I actually asked them to do a starter run to the shops for you too, to get some provisions to get you started."

Aude moved to the sleek refrigerator in the kitchen, "See, inside."

It groaned with healthy living and enough cheese and wine to throw a party.

Aude reached in, "Here," she said, "You can open it!"

It was a bottle of Dom Perignon.

"This is the way to start!" I said and looked around for some glasses.

Aude pointed, and I retrieved two tulip shaped Champagne flutes.

Pop and the Champagne was poured.

"Cheers," said Aude, "And here's to a happy life in Geneva."

Any questions?" she asked, "I usually get asked about British grocery shopping. There's Jim's British Market at Rue du Mond Ronde at St Genis Pouilly, and Intermarche St Genis (next to the Gex roundabout) has an entire aisle of British food, although some of it is quite 'quaint' like Bird's custard powder!"

"What about my neighbours?" I asked.

"Similar to you, there's another researcher, an IT consultant, someone in magazines, a banker - you know you could meet a few of them today at the party. The invitation through your door says it starts at six pm."

"And what is this party in aid of?" I asked.

"It's like I said," answered Aude, "Geneva has a public holiday today. It's called jeûne genevois which translates to 'Genevan Fast' in English. It's a historically religious holiday that is now more traditionally accompanied by eating a plum pie."

"That's why there's a picture of a purple flan on the invitation!" I said, realising it must be a plum flan.

"But no custard," said Aude.

Meet the neighbours

I went along to the party, although I was wary that they'd all be speaking French. It turned out to be hosted by an America, the Banker, named Bradley Floyd.

"C'mon in," he smiled and invited me into his apartment, which was the same shape as mine and on a lower floor. It had a comfortable lived-in look with scattered Americana around. Even a framed tee-shirt of someone from the Tar Heels and a separate picture of a ram with blue horns. Although, somehow I'd have expected a banker to have a bigger place.

"This is Jennifer Hansen, my partner; we're both from Charlotte, North Carolina and were moved out here for a couple of years. We thought we'd follow local practice when we could, so today was a good excuse for a party!"

Jennifer smiled, "Yes, hello; you must be the new neighbour that's moved in today. I'd heard you were some kind of scientist?"

I smiled back. The word sure gets around in this town,

"Well a researcher anyway, I'll be working for Brant at their R&D facility."

They seemed perfectly lovely.

An attractive dark-haired woman joined our conversation, "Hello, My name's Bérénice Charbonnier," she held out a hand, and I wasn't sure whether to shake it or kiss it.

"Let's be French, " she said smiling, and pulled me towards her lightly pecking me on each cheek.

"See, first one cheek, and then the other, You do quick cheek kisses, or *faire la bise*. It's instead of hugging in France to greet someone. Never hugs. I repeat never hugs."

Another man joined in; I instantly realised he was British," It only takes a couple of times to realise that it's the done thing here."

I asked, "So who do you do this with?"

Bérénice answered, "Family, friends, sometimes colleagues and casual acquaintances you see often. Usually men don't give each other bises unless they're family or very close."

I asked further, "Which side do you start on?"

Bradley laughed, "Who the heck knows? I don't think there's a rule, but just about all the time I start with a right cheek to right cheek."

The Brit added, "Always follow the French person's lead. They're the masters of the French greeting. If they start to turn their head and scrunch their face, they're coming at you for a bise, or cheek kiss or two."

Bérénice added, "Yes, So if they lean in for a cheek kiss, do the same. And if you wear glasses like I do sometimes, it's customary for one of you to remove your glasses so they don't clink."

The Brit spoke again, "I don't think we've been introduced, and let's skip the kisses and hugs: My name is Simon Gray and like you, I'm a researcher here. I also work for Brant, in their pharma division. I think Brant has a job lot of apartments here in this block! There's a whole group of us here. Remind me to tell you about the bus."

I turned to Bradley and Jennifer, "Thank you both for inviting me to your party, anyway, and it's a great way to get to know people!"

There was a pause as we all allowed the conversation to reset. There was gentle jazz piano playing in the background.

"So what is this special occasion?" I asked.

"I should let Père Oscar, explain - he's a priest at our local church. Both Jennifer and myself were evangelical Protestants in Charlotte, and when we came to Geneva we wanted to find something similar. Neither of us realised that Switzerland was the home of the Protestant

Reformation that swept Europe in the 1500s. Oscar, come over there, there's something you can help explain to our new friend!"

A youthful man in a Jazz Montreux tee-shirt came over. No sign of religious artefacts, unless the wrist string bracelet counted.

"Hi, I heard you all speaking in English, I'm Oscar Weissbrodt, but no, I don't live in this block! Bradley kindly invited me along for jeûne genevois! And you must be the new arrival!"

We introduced ourselves, and I realised that Bérénice's comprehensive instructions hadn't extended as far as the clergy. We both nodded politely towards one another.

Oscar continued, "Most people are unsure of why Geneva has a public holiday on a Thursday early every September."

" Jeûne genevois, or "Genevan fast," is a public holiday only in the Swiss canton of Geneva, observed on the Thursday following the first Sunday of September. Jeûne genevois dates to the 16th century and, because in English, it is often spelled jeune (without the accent) genevois, some non-French speakers think it is a holiday related to young people."

Oscar continues, "You know about the Reformation? And Jean Calvin? After he broke from the Catholic Church, he promoted the absolute sovereignty of God in salvation of the human soul from death and eternal damnation. Various Congregational, Reformed and

Presbyterian churches, which look to Calvin as the chief expositor of their beliefs, have spread throughout the world - even to North Carolina where Protestant religion thrives!" He winked across to Bradley.

Oscar continued, "Well, The Reformation did not abolish the practice of wishing to observe a penitential fast in particular situations (such as after an outbreak of the plague), or in Geneva in 1567 in remembrance of the repression against Protestants in Lyon, France, at a time when Geneva was refuge to numerous church reformateurs, and then a few years later in 1572 after the St Bartholomew's Day massacre, which spread through France."

Simon spoke, " I know about the massacre from that Alexander Dumas novel - La Reine Margot - Catherine de' Medici threw her force with a wave of Catholic mob violence, directed against the Huguenots (French Calvinist Protestants) during the French Wars of Religion.

Bérénice agreed, "Yes, there's a blood-curdling French movie as well actually. The massacre was traditionally believed to have been instigated by Queen Catherine de' Medici, the mother of King Charles IX. It took place after the wedding of the king's sister Margaret to the Protestant Henry of Navarre (the future Henry IV of France). Many of the most wealthy and prominent Huguenots had gathered in largely Catholic Paris to attend the wedding. What happened was awful."

Simon added, "Yes, around 30,000 perished in that series of massacres all over France. It also marked a turning point in the French Wars of Religion. The Huguenot political movement was crippled by the loss of many of its prominent aristocratic leaders, and many re-conversions by the rank and file. Throughout Europe, it printed on Protestant minds the conviction that Catholicism was a bloody and treacherous religion."

"Wow," I said, "It could make a serious box-set. Oh sorry, Father, you are disarming with your style."

Oscar smiled, "So now, like I think you have Guy Fawkes to celebrate the attempted blowing up of Parliament, we have this day of fasting throughout Switzerland, irrespective of the canton's confession. Geneva, being Geneva, has kept to the day Thursday rather than Jeûne Federal- a Sunday, as observed in the rest of Switzerland."

"And why plum pudding?" I asked.

Simon added, "No, it's more like a plum flan. It was one of the foods allowed to be consumed when fasting. Nowadays it's a go-to pudding here in Switzerland."

Jennifer laughed, "You Brits with your different foods. Yorkshire pudding. Mince Pies. Hot Cross Buns! It's like a whole other planet!"

I could see Bérénice looking amused.

"And Bérénice? What is it you do?" I asked.

"I work for Le genevois - The Genevan - It's a mainstream media publication. I work on the social media side of things."

Jennifer clinked a glass.

"And now for the special event!" she proclaimed.

"My friend Mr Waldo Stämpfli the chef from Le Grande Boulangerie will introduce the Zwetschgenwähe!"

With a small flourish, a kindly-looking man appeared from the kitchen. He was carrying a 60cm round flan dish, with a beautifully made plum flan inside it.

Jennifer pinged her glass again, "Father Oscar, would you say a few words,"

Oscar turned, lowered his head and said, "Bless this food and the people who prepared it. Many thanks for the meal and the company."

"Amen," said Bradley.

"Amen," said everyone else. I thought it was easy living with the Swiss language so far because everyone seemed to speak English.

Day one in Brant

Monday morning and Simon Gray had told me about the bus, which was operated by Brant and ran from Bel-Air, which was about five minute's walk from the apartment, directly into the campus. I noticed it ran along the same route as the trams for the first part of the journey and several people on the bus said there were other routes to come back outside of main commuter time.

Simon showed me the way to the pickup-point on the first day and made the expected jokes about lunch boxes but kindly introduced me to several of the other travellers on the bus going in toward Brant.

I asked them all about driving to work, but most of the city centre dwellers didn't even have cars and would hire them or even bicycles when needed. That had been my approach when I'd lived in London and the same would work here.

When I passed through the gates into the Brant complex, I could see it was vast campus. It looked almost new, and it was clear that a bus was needed inside to travel between buildings. Someone told me Brant had bought up farmland and then created a vast infrastructure plan. I had to go to the main visitor reception first, to get processed before I could get to the Research area. Inside the main reception I could also see a runway and a couple of military planes were parked alongside.

Processing of me took a couple of hours. There were questionnaires, a mini-interview, a computer test and the usual photographs. They had asked me to bring my passport and they took a photograph along with my driving license.

"It is to do with the Swiss regulations," assured Allegra Kühn, the HR Principal that was looking after my admission, "we can then enrol you on the health insurance, and we can also say that you have fully understood the secrecy undertakings involved with working here."

It all seemed tighter than the Official Secrets Act but was rewarded at the end with a contactless security badge, a lanyard and a cafeteria card. I decided I'd have to try the cafeteria card as soon as possible.

"The card will let you into any of the four cafeterias on the campus, here's a map of where they are located. You can also visit the crash-bars where you can grab a coffee and a snack. They are operated by Starbucks, but there's a special discounted pricing if you use the card."

Now this was useful information.

"Does it work with Dominoes?" I asked, in an attempt at

humour.

Allegra answered, "Oh yes, we've got arrangements for a Brant discount with several of the delivery firms in Geneva. Any amount owed is simply deducted from your next pay cheque.

"There's an online user guide and an App to show what you can do with the CafeCard. Oh, and you can use it to buy evening classes too, so if you want to learn French, this is a good way to do it. It's discounted too, although Brant subsidises the learning, so the teachers are not out of pocket."

Allegra paused, "I've called through to your department. They are sending someone over to collect you."

As she spoke, a woman entered the reception wearing a lab coat. "Ah, here you are! Your new person for the day!"

"Hello, I'm Amy van der Leiden" she said, glanced towards Allegra and then turned to me for faire la bise! And I knew what to do this time. First one cheek and then the other. But wait, then the first cheek again.

"Dutch three kiss rule, " said Amy, "I know we are in Switzerland, but I'll maintain my Dutch standards!"

"This is all so confusing, " I said, "More so than all the security procedures here inside Brant, and I'm not sure what Americans would make of all of this with their stringent Californian harassment laws?"

"The other useful thing to know is 'Pah!' " explained

Amy, "We and the French use it often."

"Thank you Ms Kühn, I'll take it from here," Amy turned and started walking back towards the entrance to the Reception area.

"Are you coming, then?" she asked as I started to trail along after her.

"We'll be introducing you to the head man in a few minutes; you know you are joining the Cyclone headgear development team? It's part of the RightMind Programme."

I shook my head, "no I didn't ever get told anything more than I was joining robotic research,"

Okay, you'll be meeting Kjeld Nikolajsen. He's Danish, in case you can't work that out, from Copenhagen actually. Very smart and quite staccato. Assume he knows all about you and think of a good question to ask him. Oh, you'll be working with me in the Research Team. I run the Cyclone development unit."

I'd just met my new team leader and now I was to meet the boss. Things were moving along.

Kjeld Nikolajsen

The route to the Research Department was along a vast internal glass domed room. There were tall trees growing and what appeared to be a woodland stream. Off to either side were glass partitioned offices and halfway up the wall was another floor which seemed to mirror the layout of the ground.

They had given me a small handbook when I arrived in the reception, as well as an App for my phone. As far as I could make out, the building was cut into three sections, each with an overarching glass canopy and a pleasantly themed walkway. Between each of the three sections was a cafe area and inside the taller end of the glass dome was a building that stretched up to four stories.

This was also just one of five large buildings on the campus and all along one side of it was access to the airstrip. I could see mainly helicopters coming and going, and hear the occasional drawn out rumble of a turbo prop taking off. I guess the building was based upon the design of an airport terminal, but it certainly had some serious money ploughed into it.

"This is one of the more impressive buildings, explained Amy, "They bring visitors here and upstairs are a few of the executive offices like the one we are meeting Kjeld in. They get a good view of the flight movements from upstairs. There's also some attractive dining rooms on a whole floor and the facility here is based upon hot-desking, even of the most senior people."

We stopped at the next gap between the glass domed buildings, and Amy looked towards an elevator.

"We could use the stairs, but I think we'll get lost if we don't follow the correct visitor route," she explained.

Sure enough, we were soon upstairs and ahead of us was a large, curved reception desk.

"Hello," said Amy, in English, "We are here to see Kjeld Nikolajsen." She pronounced Kjeld's name emphasising the 'J's is a strong Dutch manner.

"Certainly, said the receptionist. "My scanners tell me you are Amy van der Leiden and Matt Nicholson. He is expecting you. Let me show you the way."

She walked out from behind her desk and led the two of us to an office. EL6, it said on the door.

"It's okay, you can go in," she said.

Amy walked in first and I followed," Hello Kjeld," she said, beaming towards him, "I've brought Matt Nicholson, to see you."

"Well, Hello Mr Nicholson, and Welcome to Brant! - Welcome to Switzerland actually, I guess you've not had much time to look around?"

"Not really," I replied, "Although I did get to a jeûne genevois party last Thursday where I learned about Zwetschgenwähe and faire la bise, although it didn't prepare me for the Dutch variation."

"Good to hear you are getting integrated, although I think cake before kisses is probably the wrong order," quipped Kjeld. Then his face turned serious.

"You'll have Amy as your team leader as we try to develop improvements to headgear to be used in HCI. The project name is RightMind, or as the Swiss called it Esprit droit. There's all manner of implementations from Healthcare, through control systems to military if we can get this right. I'm told you have a flair for thinking outside the box, so we'll expect some significant innovation from you. Amy has an excellent team and they will be able to help you build just about anything you can envisage."

"I'll need to get the science part right though?" I added
   "Yes, naturally,"

"I was going to ask you a question..." I ventured.

"Yes?" answered Kjeld.

"Well, we've moved the technology of this through smaller animals and are now approaching the point where some human interactions become possible. I wondered about the ethics if this?"

"You have to consider the greater good," answered Kjeld, "Take cat bombs as an example."

"Huh?" I responded.

Kjeld continued, "Cat bombs. In the 16th century, they considered cats and other pets a way of bombing cities under siege. The idea was to put slow fused incendiary devices onto captured cats and birds before releasing them to return to their homes inside the besieged city. The theory was that the cat, being unnerved by the fizzing incendiary, would bolt for home through gaps too small for human besiegers to get through (and in the case of birds, over the city walls or castle ramparts)."

"There are several illustrations of the 'rocket cats' in a 1530s manuscript at the Universitätsbibliothek Heidelberg, in the 1584 Fire Book where it describes 'Make a small sack like a fire-arrow if you would like to attack a town or castle, capture a cat from that place and bind the fire sack to the cat's back, ignite the fuse, let it glow well and then let the cat go so that it runs to the nearby castle or town, and in its terror it runs to hide in a barn where the hay or straw it will catch fire.' It's more like an arson attack, rather than an explosive device, not that the poor cat would be interested in debating the difference."

"But is this real, or fanciful?" I asked.

"Is it a small loss to achieve a great gain?" asked Kjeld,

"That becomes an ethical question."

"Or a moral one," I interjected.

"Very good," answered Kjeld, "I hope you will enjoy working for us."

Amy signalled it was time to leave. As we walked back, she said, "Good, he likes you. He was testing you with that last thing about the cats. I've not heard that one before."

Diamond Hard Blue Apples of the Moon

The transparent circles
'Round the heads of all the heroes
And confusions caused by echoes
That are not for us to see
We sit on magic carpets
Its circles may be seeking for the
Diamond hard blue apples of the moon

To a blind man in a dark room
On their magic flying carpets
They must search to find a black cat
And two apples make a pair
But they are not really there,
But they're still tryin' a make a pair of
Diamond hard blue apples of the moon

Keith Emerson / Lee Jackson / The Nice

Amy van der Leiden

"We've a couple of presentations today, in the main auditorium," said Amy, "The first, which we've missed, was from a British MP, Duncan Melship. His one was about societal impacts of AI. Things like smart traffic, smart cities, smart diagnostics. All fairly 'Tin-Tin in Space' predictable. Crystalline structured hyper-cities of the future. "

"The main session is from the University of Zurich, a Professor Doctor Andreas Türkirchen, who works in their Brain Research Institute. It's aimed at post-graduate students but should still have something of interest to you. You'll be able to hear the questions that get asked too. Consider it part of reading into the role."

We moved to the Auditorium, which was an impressive lecture structure with maybe 1000 seats, flat screen displays and individual sound systems which could tune into a translated version of the lecture.

Türkirchen decided to present in English, and had a very strong command of the language, although he spoke with an American accent.

"The era of merging our minds with technology has begun. Already, we can hack the brain to treat diseases such as Parkinson's or help paralysed people move again. But what if you could install a chip in your head that would not only fix any health issues, but could amp up your brainpower — would you remember every word said during a meeting, finish crossword puzzles faster, drive better thanks to enhanced senses, or pick up a new language before your next trip?"

Türkirchen projected a screen showing a schematic of the fibres linking inside of the brain. It was a wispy yellow creation and reminded me of the kind yellow dyed wig that Donald Trump might wear.

"You who work here already know about the developments in 'neural lace' - the sci-fi inspired interface designed to blend human brains with computers. It's the idea to keep pace with ever-smarter artificial intelligence through the use of implants that will make us smarter."

I remembered that Elon Musk had already been using one of his new companies to investigate neural lace, although it struck me as a giant leap from inventing the concept to somehow grafting it into someone's' head. It would need to be exclusive-hotel housekeeper bed-making perfect in the way they smoothed it across a brain surface.

"Under any rate of advancement in AI, it will leave humanity behind by a lot. The benign situation with ultra-intelligent AI is that we would be so far below in intelligence, like a pet maybe."

I remembered the previous analogies here. Like cat mathematics: Cats might be smart enough to know how far they could safely fall, but they'd still get lost with counting more than half a dozen kittens. I'd studied cat brains briefly, because of their similarity to humans, even down to the folded surface area, although the hippocampus, amygdala and frontal lobes were only 3 to 3.5% of the total brain in cats, compared to about 25% in humans. The frontal lobe is the action cortex and controls numeracy, language and decision making. Cats have an altogether more reflexive existence and it's no co-incidence that 'copycat' is an idea. It's how cats survive.

Türkirchen was continuing, "But consider carefully, the risks of incorporating neural lace far outweigh the gains at present. A device capable of replacing a smart phone interface isn't enough to win any technological races."

He flipped up some further diagrams. They were of neural lace. I noticed immediately that they were all CGI renderings except a picture of a young girl wearing a bathing cap styled head sensor probe.

I could see where this was leading. It was a micro-scaled mesh similar to my old primitive woven-metal AM antenna and band-aid solution. This would be a conductive mesh, which could be injected into the skull and would then be spread on the brain and also onto the outer skin surface of the head. Imagine a sandwich with mesh outside, the skull in the middle and then the mesh again on the surface of the brain. A grid of co-ordinates that could be individually accessed. I could see problems, but the approach had some potential. But my own thinking had already moved beyond this risky approach.

Türkirchen continued. He was admitting it now, "And we know that Super-smart AI isn't right around the corner, and the goal of creating brain implants for healthy people in just eight to 10 years might not be realistic. But scientists are hard at work on technology that could boost our mental skills. Some of these gadgets could be worn right on the skin, but the most powerful ones will be nestled in the brain. Here's how they might work."

That was a challenge, because the first objectives seemed to be to make something that could provide rudimentary control for a smartphone. I could see it for accessibility, but it wasn't the most saleable:

'Let us drill your head and you'll be able to run your smartphone from thought waves.'

"The first steps towards using computers to upgrade our mental skills will be improving the connection between brains and machines. Initial developments will beef up our senses and allow us to control robots from a distance. Technology that can use the electrical signals crackling through our nervous systems to help command computers already exists in some form. People with severe paralysis can use brain-computer interfaces to control a cursor on a screen. Others have been able to move robotic limbs or even to fly planes."

Although the flight of unmanned planes can be as easily accomplished by hand-operated joysticks, like some advanced Playstation simulation.

"Researchers can also use the technology to deliver messages to the brain. By sending an electric current into the correct neurons, scientists have been able to restore a person's sense of touch or hearing, treat tremors caused by Parkinson's, or send very simple signals from one brain to another."

That's more like it. HCCH interaction. Human-Computer-Computer-Human interaction. With the middle computers providing some form of interpretive moderation.

"Researchers are now exploring whether these technologies could also sharpen certain cognitive skills. One non-invasive technique called transcranial direct current stimulation works by sending electricity through the scalp. Some scientists (and DIY brain hackers) hope it can help improve skills like learning and memory. But it's not clear yet if this brain-zapping technology is effective."

That's a reference to my olden hacker days, with the PP3 battery, the wire mesh and a couple of Band-Aids. I couldn't help but smile.

"Meanwhile, the Defense Advanced Research Projects Agency (DARPA) is investigating a less direct approach: sending electrical pulses into the body. Research indicates that zapping certain peripheral nerves — which connect the brain and spinal cord to the body — may help people learn skills faster. The most promising target for this approach is the vagus nerve, which passes through the neck. It is like tapping into the information superhighway carrying information from the body to the brain,"

I was thinking that he was proposing a part chemical and part digital approach. The only thing was, it needed a slash to the neck like something out of a bad Mafia movie.

Türkirchen continued, "Sending electricity through the vagus and other nerves may prompt the brain to release chemicals that alter connections between different neurons. This is already a key part of our learning, but by using a machine to rev up this natural process, people might be able to "tune" their brains to recognise important details with less practice."

I knew it. Using chemicals to accelerate or stimulate learning. It's a more random approach than desirable. It was a path I'd seen the Russians walk along.

Türkirchen continues, "Implanting electrodes inside a human brain is still risky, so it's only done to treat neurological diseases. However, non-invasive technology can't zero in on the exact neurons it would need to stimulate to boost someone's mental skills. For that, scientists will need to come up with technologies that can be embedded inside our skulls."

That's why my Band-Aid patches were only useful for general cross skull stimulation. I knew back then they could not be so precise.

Türkirchen put up another couple of diagrams: "At that embedded level you gain access to the actual source code of the brain — our neurons that are firing. That's the entry point where you gain the highest potential of what you can do."

The diagram showed neutrons terminated with sensors which, in turn were interfaced to a chip. Good in theory, but very difficult to achieve in practice. I guess that's why he was hinting at the neural lace idea. Fifty sensors into the brain and surely some of them would prove usable. It's still an incredibly reckless hit-and-miss form of experimentation.

Türkirchen continues, "How these surgeries might become outpatient procedures remains to be seen. Eventually, it could be automated like laser eye surgery. But whatever we put inside the brain; it will have to be more sophisticated than today's technology. At first, it might look a bit like today's brain-computer interfaces. Many, such as BrainGate, use a pad with dozens of needle-like electrodes to plug into the brain. It looks like a mini hairbrush. We might adapt this design for implants that would be slipped into different areas beneath the skull."

He added, "Probably it would be many implants first, aiming towards a single implant with many electrodes," These implants might initially be the size of a smartwatch."

Smartwatch - I liked the idea of repurposing the smartwatch computer. However, and it's my idea, put it into a discrete backpack and it could be the size of a smartphone. Then you'd have some significant processing power, but I'd hang it from the spine, not the head. Like the way those Mafia movie guys carry their pistols down the back of their trousers.

Then he brought in the idea of nano-sensors: "Or we might take another approach, like injecting nano-sensors that would be deployed throughout the brain and controlled by magnets or radio signals. Some scientists are looking into out-of-the-box interfaces made from electrode arrays printed in a tattoo."

No wonder the skull is such a good protective box for the brain. To try to stop humanity from fiddling around with it. I'm surprised it hasn't got one of those 'High Voltage - do not Tamper' warnings on it.

Türkirchen added, "Right now we're really limited by the ways we're recording and stimulating cells in the brain. A brain implant might work by sending pulses of electricity directly into neurons that help govern a particular skill although how this will boost performance isn't clear. Or the chip might be a more powerful version of DARPA's technology that would tinker with the chemicals the brain uses to help itself learn. Or the chip might give the brain some extra memory and processing power, so recollections and knowledge could be stored externally rather than forgotten. 'Downloading' memories directly into the brain is definitely a long way off,"

And he said, "There's no one area of the brain an implant could zap to make us more intelligent across the board. We'll have to target specific skills. Abilities that rely on the senses (like picking up a new language) will be the most straightforward to enhance. We could also sharpen our focus relatively easily."

Türkirchen had run out of things to say. He wanted to demonstrate the Science was flummoxed: "Eventually, we could target other abilities, like reasoning, creativity, willpower, or judgment. However, these skills are powered by many brain areas working together, so figuring out how to improve them will take longer."

Flummoxed. There, he'd said it. They still don't know how to do it. But they certainly know how to dress it up as plausible.

Then for the science fiction part: "So what happens if we do merge with machines — or at least use them to make ourselves smarter? Who would be able to use cognitive enhancement technology? How much might it cost? How would we make sure the technology feeding into our minds stayed private and secure?"

I was thinking about Johnny Mnemonic, the story by William Gibson, later a film with Keanu Reeves. It showed a dystopian view of the future where Johnny has a cybernetic brain implant designed to store information in a world dominated by megacorporations. I'm sure Türkirchen must have seen or read it.

He continued, "We can't fully imagine how we might use a new tool until it's in our hands (or brains). No one could have guessed all the ways we would use the Internet, or the knowledge we would gain from all the books made available by the invention of the printing press.

"In the meantime, we have plenty to appreciate about our brains. Sure, the human mind has its limitations; there's only so much we can learn and remember. But we have the brain, with its billions of information-ferrying neurons, to thank for our ability to read a book, learn a recipe, feel compassion for others, navigate websites — and dream up technology to make us even cleverer."It was like a Disney wrap-up at the Epcot Center. It just needed a deeper voice to make it more special.

"The brain is the master tool and the master of all tools, everything we are, everything we do, everything we're capable of is because of our brain."

I think he missed a trick by not having some close-out music at the end. There was a ripple of applause from around the auditorium, but I sensed that many of the people there must have been thinking similar thoughts to me. I thought I'd try it on Amy.

"What did you think?" I asked her.

"It shows what we are up against," she replied, "All the pressures from the funders - that's Brant and the US DoD, yet we're up against some practical limits."

"Explain?" I asked.

"Well, there's the moral one of opening up perfectly well-functioning humans, to drop in brain boosters. And there's the one that the boosts may not be capable of much for the next ten years," answered Amy, "So Matt, what did you think?"

"Well," I answered, "There's a chance that everyone is going about this the wrong way!"

Amy squinted her eyes at me, "They said you were either

brilliant or intensely arrogant but couldn't work out which. I guess we'll find out over the next few days. Come on, I'll introduce you to the team."

Evening in the bar in Geneva

I was amazed at how quickly the first day had passed. By the time I'd been shown back to the Lab, allocated my workstation position, met the IT Guy and some of the rest of Amy's team, it was time to go home again.

I was quite glad it was the end of the first day. I'd loaded myself up with so many new images and thoughts and needed time to relax and process everything.

One of the team approached me. "Mr Matt?" he asked, "Can I give you a ride back into Geneva. Mr von Westendorf and I usually go for a drink after work Monday evening, and we'd like to invite you along."

"Wow. That would be great," I answered, "Thank you," I worked out that they were both native German speakers and that Schmiddi was really called Doktor Schmidt. I reasoned that von Westendorf was probably Dr. Von Westendorf too.

Schmidt gestured to von Westendorf, who walked over. Schmidt was a shorter and more rotund man, wearing a suit and looking quite happy. By comparison, von Westendorf was a taller man, dark curly hair, glasses and a small moustache. His resting face was more sombre.

We made our way to the parking lot and into Schmiddi's enormous blue Renault car. He started on the route back to Geneva, at an alarmingly high speed. I was sitting in the back, with von Westendorf riding shotgun in the front.

Schmiddi started, "Yes, we are both from Germany - me from Baden-Württemberg and Dr von Westendorf from Bayern - that's Bavaria - as you English call it. And you are from where?"

"Originally London, although I've spent some time in Cork, Ireland as well."

"Excellent, we practice our English with you. We are trying to learn English the way she is spoken in London rather than in Washington. It is so difficult because we have so many American programmes on television and I don't think Downton Abbey is a good representation of English."

"No, you are probably right, although it's amazing how many people around the world have seen the show!" I answered.

Dr von Westendorf said, "When I was in London, I visited the place that they use for Downton Abbey. It is called Highclere Castle and is outside of London around 100 kilometres.They told me there was a waiting list to get in, it was so busy! I explained to them that I was visiting from München, especially to see the Highclere and they let me come along. It was fascinating - the original owner was one of the people who discovered Tutankhamen's tomb, in Egypt. I guess they could have added that into the Downton story somehow.

"And I realise your name is Herr Nicholson, now I have seen it on your badge. I know in England you are less formal with names, so if it is alright, I will call you Matt. You may call me Hermann and Dr von Westendorf ist Rolf."

We were back in Geneva now. It was a startlingly quicker journey than on the bus in the morning.

"You are staying in the Apartments near to the Bel-Air bus-stop?" asked Schmiddi.

"Yes," I answered,

Schmiddi replied, "Okay, most of the bars around there are linked to hotels. We'll go to one of them tonight, so that you can't get lost going back home."

"What about the MO?" suggested Rolf.

Hermann replied, "Good idea, we'll go to the Mandarin Oriental; they have a good bar inside. It looks like a proper bar too, rather than some of the hotels where it has been made like a cocktail lounge or an airport waiting area."

We pulled up outside and Hermann said, "You can go inside, I will park the car around the corner."

The building looked like an 80's concrete slab with a large plate class window but transformed inside into a quiet hotel lobby with bar off to the side.

"This is most kind," I said to Rolf.

"Let's find a table," he said.

We sat and a few moments later a waiter appeared with a drinks menu. He spoke French and Rolf replied in what sounded like quite elegant French. Then he switched to English, "I have ordered Hermann and myself a bier; maybe you'd like one too?"

I quickly looked at the menu and pointed to one. It was a dark Bock, brewed in Switzerland.

"They use the same Reinheitsgebot here as in Germany, the beer is made along German lines in Switzerland," explained Rolf.

Hermann appeared, "I'm parked underneath in their car-park," he explained, "We don't usually come to this bar."

The three beers arrived, two pale beers in slender glasses and mine in a glass more like an English 'tall' glass.

"Yours is a half," said Hermann, "That's half a litre or almost one of your British pints. But be careful. It is easy for the British mind to play tricks with the quantities."

We made some small talk and about how we'd each come to be working for Brant. Money played a significant part for both Hermann and Rolf. They had come through an agency and had engineered a special taxation arrangement.

Then they asked me about my specialisation. I told them it was robotics and the Human to Computer interface. They were both very interested because it was something they had worked with for several years.

"The brain is so well protected and dangerous to interfere with, " said Hermann. We have, naturlich, a few experiments with animals and it is now possible to place electrodes very precisely, but there is still inevitable outrage at anyone considering doing this to a normally healthy person."

"But what is the specific pressure from Brant?" I asked.

"Nyah, it is an arms race, really," said Hermann.

"Exactly," continued Rolf, "They want to develop augmented humans for battlefield purposes."

"What? Like supermen?" I asked, incredulous.

"No, nothing like the movies. Far more mundane. They want to provide logistical support from humans. The amount they would need to carry would be dramatically reduced and the on-battlefield telemetry they could handle could be ten or one hundred-fold what we see today."

"But wouldn't it apply all across NATO?" I asked.

Rolf explained, "Yes, but that's not the point. It is not about military advantage. It is about financial reward. The first mover in this marketplace can make a lot of money. That is why Brant, with its military contractor background, will want to be seen to have developed the early models."

Hermann added, "It's also why Amy is under intense pressure from Kjeld Nikolajsen to get something prototyped. Hey, English Boy, another drink?"

I nodded agreement, and they ordered another round of the same.

"So, what's 'Same again?' " I asked.

"la même chose !" said Hermann.

"Or encore le même!" added Rolf, "You are learning French quickly."

"Okay, and another important topic," said Rolf.

"Which football team do you support? Mine is Football-Club Bayern München! - Stern des Südens (Star of the South)!"

"La même chose?" queried Hermann.

I nodded, but remembered I'd have to be careful with this pure-German brewed-in-Switzerland beer. But, in actuality, I felt absolutely fine on it.

Avoir la gueule de bois

Tuesday morning and I realised Monday's beer was really quite strong.

I had been at the office for one day and had completely smashed myself into the ground. The hangover was like someone has found a large pickaxe and was working their way around my head.

I remembered all the advice about congeners and dark drinks being worse than light drinks. At this rate, I wouldn't be able to get to the office. It would look terrible if I didn't make it, although I guessed if I did go in then people would tell immediately.

I decided to pick up my ashen-faced self and get along to the bus stop.

Simon Gray was already there.

"Mate!" he said, "What have they done to you? You look awful?"

"I was out with Schmiddi and von Westendorf yesterday. We had a few beers."

"Oh Matt," he said, "Those two have a bit of a reputation, you know, as boozers. Look let me see,"

He rummaged in his bag. "Neurofen Plus; they should shift it," he said, "But look, just because I work in Big Pharma doesn't make me an expert."

"Oh, thanks," I said, as I swallowed down the tablets.

Simon snapped off another two still in the tinfoil.

"Put these in your pocket for around midday. But beware, these are Plus versions, so they also contain codeine. It's banned in America."

"I can't really get any worse," I sighed.

The bus arrived and we climbed on. I looked out of the window and felt the waves of Neurofen Plus balming my soul. Simon thoughtfully kept quiet and was reading an English paperback.

Then I was in the office. I caught Herr Schmidt out of the corner of my eye. I decided it was appropriate to thank him for the previous evening and walked across.

"Thank you, Herr Schmidt, that was a most entertaining evening!"

"Yes and thank you for coming along; to be honest, we thought you might be the worse for wear today after drinking that many Dunkel Bocks. They are about 7%, so I'm amazed you haven't got a massive gueule de bois this morning! We were both very impressed, Englishman!"

At that moment Dr von Westendorf appeared. He was wearing dark glasses.

"Merde!" he said, "c'était toute une nuit! - That was quite a night! And you, Matt, are a formidable drinker!"

If only they knew. Thanks, Simon for the medication. Now I was scheduled to meet Juliette Häberli, one of the research scientists in our team.

Cat learning

Tuesday and I was in the lab, learning about the experiments conducted recently by the team. An attractive and softly spoke Swiss team member, Juliette Häberli was explaining the experiments. I could sense that she had some of my misgivings about the control of living creatures and the implications for humankind.

Juliette began, "We started with cats, they are bigger to work with, which makes some things easier than working in the micro measurements need with rats. Of course, we had a range of non-invasive experiments to begin with."

It implied to me that the cat experiments would get more intrusive as they developed.

I noticed Juliette's movements were economical and measured. I wondered whether all this cat talk was getting to me. She continued, "We conducted some key experiments on cats' learning capacity, based upon the work of Edward Thorndike. In one experiment, cats were placed in various boxes approximately (51 cm × 38 cm ×

30 cm) with a door opened by pulling a weight attached to it. The cats were observed to free themselves from the boxes by trial and error with accidental success."

I said, "It backs up my own thoughts that cats are really not that bright. They don't exactly learn unless they can see a reward at the end of it. And they learn to hunt and so on from their early play."

Juliette looked a little surprised at my assertions; I caught the smallest of changes to her expression at the corner of her eyes, "Though cats did perform worse on occasion, we found that as cats continued the trials, the time taken to escape the boxes decreased in most cases. We determined that the cat followed the law of effect, which states that responses followed by satisfaction (such as a reward) become more likely responses to the same stimulus in the future."

I said, "Yes, but that is my point, they are reward motivated!" Romanticism aside, I was sceptical of the presence of intelligence in cats and saw contemporary write-ups of the sentience of cats as partisan. Deducing from facts and more especially in the choice of facts for investigation. Bending reality to fit the case.

Juliette was looking at me like she was playing with a fluffy ball of wool.

"Then we tried the observational learning experiment with kittens. The kittens that were able to observe their mothers performing an experimentally organised act were able to perform the same act sooner than kittens that had observed a non-related adult cat, and sooner than the ones who, being placed in trial-and-error conditions, observed no other cat performing the act."

She twisted delicately in her seat, "We'd previously received wisdom that cats need to watch their own mother in weeks two to eight to learn some key cat skills. It is the same period that kittens become habituated to humans and lessen their fear of them."

And something that made me pause to think, "Cats will occasionally reach up with a paw to put a person's hand on the exact spot where the cat wants to be scratched or stretch up to indicate they want a door opened that they cannot reach or turn without opposable thumbs. These behaviours they would likely have to learn through observation, social interaction, and accumulated experience. Even without being trained by humans specifically to perform these tasks, these behaviours have mainly been discussed anecdotally, and it would need more studies before researchers can conclude how developed the cats' theory of mind may be."

'Anecdotally' stuck in my mind. So it was still an inexact science, this study of cats.

Juliette added, "The term 'copycat' comes from a popular belief that cats, in an extension of their instinctual tendency to copy each other's stalking, pouncing, chasing, and wrestling to improve their hunting skills, will follow any sufficiently enticing movement by a familiar cat, human, or other accustomed nearby nonthreatening animal. Theory of mind may be helpful for them to be able to learn the potential evasive actions of different types of prey animals, without automatically engaging empathy for what they are about to kill."

I said to Juliette, "I thought that cats and dogs had learned to become domesticated, so that they would not become an automatic choice for food in ancient times?"

Juliette smiled at me; I felt the room light up. She continued," Yes, domesticated cats are the basis of most cat intelligence studies. The process of domestication has allowed for closer observation of cat behaviour and in the increased incidence of interspecies communication, and the inherent plasticity of the cat's brain has become apparent as the number of studies in this have increased scientific insight."

She paused to check that I was taking it in. I noticed the small diamond studs in her ears and the small tattoo in the inside of her wrist. The tiny blue shape looked like a scarab beetle.

"Hey, can I ask you something personal?"

Juliette looked at me, "I guess you can ask," she said.

"What is that tiny tattoo? Is it an Egyptian scarab?" I asked.

"Yes, you are almost right. It is the Hercules Beetle. But drawn very small. It is a symbol of strength. I practice karate and it is sometimes helpful. Especially when I am up against men who say things like 'You fight good for a girl.' I let them see the beetle when I have them trapped."

"You have a belt?"

"Everyone has a belt. I am Renshi. It means 'polished expert' "

I straightened in my chair. I would need to pay close attention to Juliette. I didn't really want a close-up of the beetle. Not like that, anyway.

She continued, " Changes in the genetic structure of a number of cats have been identified because of both domestication practises and the activity of breeding, so that the species has undergone genetic evolutionary change due also to human selection.

Cats' intelligence may have increased during their semi-domestication: urban living may have provided an enriched and stimulating environment requiring novel adaptive behaviours. Such scavenging behaviour would only have produced slow changes in evolutionary terms, but such changes would have been comparable to the changes to the brain of early primitive hominids who co-existed with primitive cats.

"So, we think the humans drove the evolution of cats, on their own terms?" I asked.

"Almost certainly," answered Juliette, "And they speeded up the evolution too."

"That's amazing," I said, "I had suppositions on much of this, but you have reinforced it with research and facts."

"Are you saying you knew all this already?" asked Juliette, a short frown breaking out around her eyes.

"No, I didn't. I meant to say I'm impressed at how much science-backed research you have done. And you seem to have done this non-invasively too."

"Yes, I hope we can find some non-invasive procedures," she smiled again and looked me in the eyes. Then I realised, she had trapped me; I was smitten. That's why she had let me see her tattoo.

"This evening?" I asked, "Can you explain some of Geneva to me the way you explain about the cats?"

"Ce n'est pas beaucoup d'incitation, mais d'accord, je peux vous apprendre à être plus suisse," she said, "Not the greatest inducement, but I can show you how to be more Swiss!"

She smiled again and I suddenly felt hot under the collar. I wasn't sure if it was the Neurofen working.

I think

First Man: I think, I think I am, I think.

Establishment: Of course you are my bright little star.
I've miles
And miles
Of files
Pretty files of your forefather's fruit
And now to suit our
Great computer
You're magnetic ink.

First Man: I'm more than that, I know I am, at least, I think I
must be.

Inner Man: There you go, man, keep as cool as you can.
Face piles
And piles
Of trials
With smiles
It riles them to believe
That you perceive
The web they weave
And keep on thinking free.

In the Beginning - Graeme Edge

Juliette's Theory of Mind

The rest of Tuesday spun past. They had given me an email account and access to various servers, along with a long list of Research Papers to read. I wondered when they would expect to get me started on something, but I suppose it was only Day Two.

I was about to finish for the evening when my desk phone rang. It was Juliette.

"I'm at the entrance, are you coming along?" she asked.

"Yes," I answered and hurried along the walkways to meet her.

She was another one who drove to the office, and we walked down into the car park, where she blipped her key and a white Porsche responded.

"Nice car," I said,

"Thank you, I got it with the bonus payment from Brant - it's not new though."

I climbed in and couldn't tell how old it was. It certainly looked pristine.

Then Juliette quietly moved the car out of the carpark and onto the main road. She didn't drive like Schmiddi the previous day. Her's was a more controlled drive, with signalling and courtesy.

"I hear you were out with Schmiddi and von Westendorf yesterday? - I'm surprised you could walk this morning, let alone take in one of my lectures about cats."

"I'll come clean, " I said, "I was heavily medicated with ibuprofen and codeine this morning."

"Codeine," she said, "An opiate narcotic. You'd need a prescription to get that here in Switzerland."

"It's ironic considering how many drugs the Swiss manufacture,"

"Yes, but because it's made here doesn't mean you should be able to consume it."

I noticed her using the gearbox paddles like a racing driver as she overtook a small string of slow-moving cars following a tractor.

I was also aware of just how many fancy-looking Porches there were on this Swiss road. It was like being back in London.

"I thought we could go to somewhere on the lakeside,"

said Juliette, "There's a good place on the north side of the Lake. It isn't too far outside Geneva but is better than going to the regular city centre places.

We were soon back in the centre and then started out along the other side of the lake. We soon came to Quai Wilson and Juliette guided the car into a parking spot close to the lake.

"Here we are, La Terrasse by Dominique Gauthier, it's one of my favourite chill-out places in Geneva."

I looked at the water jet nearby across the lake and the distant mountains. It was early evening and there were small lights and the ting and plink of miscellaneous yachts moored on the jetties. A promenade separated the pedestrians from the still busy traffic.

"Magical," I said, remembering this was two work colleagues meeting, not a date. Stay professional.

"I know," agreed Juliette, "It's a tough place to beat."

We ordered some drinks - mercifully Juliette showed me the less potent forms of beer and she selected a sparkling water.

"So why did you come here?" she asked, "Do you know some things, or are you here to learn?"

"I was head-hunted," I explained, "Bob Ranzino of Brant interviewed me and then asked me to come out here. I was just at the end of a relationship, and so it made sense for me to make a move. Learn some new things and hopefully have something to offer."

Juliette smiled, "I did a quick google of you," she said, "But there wasn't much to see. Nowadays that is either because you are technophobic or else that you are very techno-literate."

"I think it is probably the latter," I replied, "I've done some things with technology in the past, and its better that they stay there."

"Cyber currency, I think?" asked Juliette, still smiling.

"Something along those lines," I replied.

"I looked in Brant's own files and found out more about you there than in the whole internet. You created a currency mining system," it says.

"Yes, that's right, but like I said, it was a long time ago."

"But it seemed to raise the interest of some governments?" continued Juliette, "Don't worry, I will not tell everyone, I'm just interested in this new team member."

"You seem well informed. You will need to show me where you got this information,"

"I can, although I think you'll need to have been here for at least your probation cycle before they will give you access."

A server appeared and asked us if we would like to order

some food. I glanced at the menu and noticed that the entrees were around CHF 60-90, which was around US$65-$100 per person.

"We can have Salade de poulpe caramélisé au piment d'Espelette, lentilles de Genève and Les desserts à partager, if you like - to share, " she suggested.

"Hmm, caramelised octopus salad, with lentils," I searched around for my schoolboy French, "If you recommend it, then I'll go for it!"

"Très bien!" she ordered the food, and we continued with our drinks. A moment of silence.

"You are very kind to show me Geneva like this," I ventured.

"My pleasure to do so; I could see you were - how you say - falling for me in the laboratory earlier today. But I thought we could reset that idea as we work so close together. It is much simpler for us to be good friends than to start something else. I could fall for you too, but then where would we be? A source of office gossip and a fragile relationship. Now I know you have just split with someone else; I don't want to be the rebound girlfriend either. This way we clean the air and can be great friends."

Okay. Juliette was plain speaking. And I suppose five days into my Swiss adventure was also pushing it somewhat.

"I have plenty of single friends, too, when the right time

comes along," she said, "And you will need a Swiss girlfriend later. It will be the best way to learn the language, English Boy," she smiled, and I was realising that the lab people probably had a nickname for me. Schmiddi had called me 'English Boy' yesterday, now Juliette. I would have to ask.

"So do I have a nickname in the office?"

"Non! It is too soon for you to be getting a nickname. Schmiddi's name was earned as was Westi's - I expect yours will become Nikki, but only once you have earned it. Or Matti."

This little adventure was moving from magical to me getting a gentle ribbing from Juliette.

"So, you've checked me out - what about you then? Can you give me some of your story?"

"I started as a psychologist. I studied here in Switzerland, but not in Geneva. I studied in Zurich and gained my Diploma there."

"But doesn't Zurich speak German?"

"Yes, but we Swiss are often multi-lingual. I speak French, German, English and oh, some Italian. You could say I'm about equal in the first three languages, although I know my English is sometimes a little American because of all the television shows."

"After I'd finished my time in University, I joined a Psychiatry practice and soon met many people with all kinds of problems and challenges. One of them was a senior guy at Brant - Levi Spillmann, who transferred to Brant bringing with him a product called Createl. I gave him some assistance to get back on track after a breakdown. It is because of my time in practice that I realise how vulnerable men fall for me. That's probably why I have such a good line in self-defence. It's really for the best."

I smiled, and she still seemed to give me the deep-eyed treatment. The arrival of the salad interrupted my thoughts.

"Hmm - great choice," I said as we both looked at the decidedly shareable dish.

"Well, then Levi said that Brant was looking for people to join their HCI Practice and that I struck him as a suitable candidate. He worked on the defence systems research. I wasn't interested in Brant to begin with, but he told me more about it, explained it wasn't all defence work and also the salary involved, and it really made me reconsider my plans.

"I'd been with a boy - Jacques - for the last two years as well, but it wasn't going anywhere and we both knew it. There needed to be a catalyst to change everything, and this was it. I wound myself out of my relationship with Jacques, got a new apartment through Brant, and changed everything. I haven't looked back."

"Very discreet too," I said, "Is Levi still at Brant?"

"That's an astute question! But the answer is no. Prepare

to hear something gory. He was a boating nutcase and kept a yacht on the lake here. One day they spotted his boat becalmed in the middle of Lac Léman, by Thonon-les-Bains, which is right in the middle and at the widest point. They sent a rescue boat out to him, but he wasn't on board. His body washed up in Evian, France, around two weeks later. They recorded it as misadventure."

"How awful," I shuddered.

"What can you tell me about the rest of our team?" I asked.

"You will need to find that out for yourself," answered Juliette, "Let's just say we are a close-knit team and watch out for one another. They usually give Amy the tough demands by Kjeld Nikolajsen, and she acts as a deflector for the rest of us. You know. Keeps us protected. The thing to remember is that we all try not to over-promise and under-deliver. It's a quite Germanic work culture. Detailed, accurate and well-planned, but we say when we think things are going to be late. Amy and Kjeld handle the upstream manipulation of 'facts' to the American bosses."

"What to people like Bob Ranzino?" I asked.

"Yes, there's a few 'fly-by handshakes' from Corporate head office. You must know how to treat them. They are all quite smart, but relatively unsophisticated in the ways of Europeans. You'll have a natural advantage with English and your command of 'awesomeness' and understatement, which can completely bypass American brains."

"But wait, aren't you being over stereotypical?" I asked.

"Maybe!" replied Juliette enigmatically.

"So, what are you researching, then?" I asked to bring the conversation away from washed up bodies.

Juliette smiled, "Theory of Mind and its applicability to Human Computer Interfaces."

"Interesting," I said, "Bob Ranzino asked me something about that. You know, the question about whether a cyber cockroach will run from flames when commanded to walk into them."

I realised I'd accidentally tuned the conversation from washed-up bodies to cockroaches. Hardly appropriate when eating an octopus salad.

"I remember Theory of Mind as being about the assessment of an individual human's degree of capacity for empathy and understanding of others. One of the patterns of behaviour that is typically exhibited by the minds of both neurotypical and atypical people, that being the ability to attribute—to another or oneself—mental states such as beliefs, intents, desires, emotions and knowledge."

Juliette smiled; she could tell I only had rudimentary knowledge.

"Yes, you are right. And I'm looking at whether machines can possess similar attributes, or whether those attributes in an animal can override a machine - Just like the burning building scenario you referred to. Theory of mind as a personal capability is the understanding that others have beliefs, desires, intentions, and perspectives that are different from one's own."

"For a being or a machine, possessing a functional theory of mind is crucial for success in everyday human social interactions and used when analysing, judging, and inferring others' behaviours."

"I guess it is more behavioural," I hazarded in a vague hope that I could keep up.

Juliette continued; she was still looking at me intently. "You could say that. Theory of mind is distinct from the philosophy of mind. Deficits can occur in people with autism spectrum disorders, genetic-based eating disorders, schizophrenia, attention deficit hyperactivity disorder, cocaine addiction, and brain damage suffered from alcohol's neurotoxicity; although deficits associated with opiate addiction reverse after prolonged abstinence."

"Cocaine, alcohol, opiates; that's a toxic list, where chemistry has been introduced which upsets the balance?"

Juliette continued, "Now suppose we position what the mind does as an output from a process. The output such as thoughts and feelings of the mind are the only things being directly observed so the existence of a mind is inferred. It's like those old fairground attractions where you ask the puppet in a glass case (like a Sorcerer or fortune teller) a question and he spins around with an answer. The oldest fairground machines in the late-1800s used a selection of cogs and a man would sit behind the machine in a tent, listen to the questions and make up an answer. It served well as an illusion of a mind in the machine."

"I see it is like the question of what is the mind?"

Juliette continued, "Exactly. The presumption that others have a mind is termed a 'theory of mind' because each human can only intuit the existence of their own mind through introspection, and no one has direct access to the mind of another so its existence and how it works can only be inferred from observations of others."

"Mind theory based upon inference and introspection?" I asked, aware that this was getting deep, "But does that mean that a machine could also have a theory of mind?"

Juliette spoke, "We're straying into Artificial Intelligence now. Theory of mind appears to be an innate potential ability in humans that requires social and other experience over many years for its full development. Different people may develop a more, or less, effective theory of mind. Theories of cognitive development maintain that theory of mind is a by-product of a broader hyper-cognitive ability of the human mind to register, monitor, and represent its own functioning."

"Hyper-cognitive?" I asked. We'd finished the salad and the waiter was clearing the table.

Juliette looked at the waiter and then at me, "Consider

the concept of empathy, meaning the recognition and understanding of the states of mind of others, including their beliefs, desires and particularly emotions. This is often characterised as the ability to "put oneself into another's shoes". Can a machine do this? Can a machine understand the ideas behind this?"

I said, "I see, we are drifting towards Rene Descartes - you know - 'I think, therefore I am,' "

"Exactly," said Juliette, "We attribute human characteristics to pets (like Fido the dog), inanimate objects like Henry vacuum cleaners, and even natural phenomena like Old Faithful water geysers. Most car Sat Navs get given a name.

"It's like taking an 'intentional stance' toward things: we assume they have intentions, to help us predict their future behaviour. However, there is an important distinction between taking an 'intentional stance' toward something and entering a 'shared world' with it.

That's the area that the HCI must cross. Will the machine believe it is sharing the mind of the human, or will it simply piggyback to accept human commands? An intentional stance is detached, and we resort to it during interpersonal interactions. A shared world is directly perceived and its existence structures reality itself for the perceiver. A shared world is the melding of the information space between the machine and the human."

"And such a shared world could be one inhabited by lovers, or a mother rearing a child?" I asked.

"Yes," answered Juliette, "Such situations seem capable of producing many of the hallmarks of theory of mind, such as eye-contact, gaze-following, inhibitory control and intentional attributions."

"That would have some deep implications for an AI device hooked up to a human," I said, "The machine would have to love the human or treat the human as its child."

"Yes, we must find another model," said Juliette, "or else we need to make Richard Brautigan's poem come true: '*I like to think of a cybernetic ecology where we are free of our labours and joined back to nature, returned to our mammal brothers and sisters, and all watched over by machines of loving grace.*' "

Backbeat

"Can I give you a ride back into the city?" asked Juliette.

"I'm staying in an apartment near to Bel-Air," I explained, "If it is out of your way, then I can call a taxi."

"Non, that is perfect," said Juliette, "You are not in that block on Rue de la Confédération?"

"Yes, I am actually,"

"I think that must be a block owned by Brant. So many new arrivals start out there."

"The apartment seems good, to me," I observed.

"Yes, but eventually you'll want something with more space, or a better view," said Juliette, "Most people are in there for around six months before they move on to somewhere else. I am along the road back towards the campus, but my balcony looks out on the Lake."

"It sounds idyllic, maybe I can take a look sometime?"

"Naturellement!" smiled Juliette.

We climbed into her car and made the brief journey back. As I was leaving the car, I spotted Simon Gray walking towards the entrance. I waved to Juliette as she executed a neat little U-turn and was away in a matter of moments.

"Hey, Matt! Was that Juliette Häberli?" he asked.

"Yes, we were out for a drink."

"Punching above your weight?" asked Simon.

"No, it was a lovely evening, we were simply two co-workers talking about Research!"

"I'm going to put you down as a 'Cheval Noir' !" said Simon, " I guess you won't have had time to find out all about the hot Ms Häberli?"

Then he added, "You know what? Come on, let's have a drink in mine, I can fill you in on the detail."

We took the elevator to his floor. He seemed to have another identical apartment to mine, but this time kitted out with British paraphernalia. Bottles of British beer in a row on the windowsill, it reminded me of a student lifestyle.

Simon cracked open a couple of beers, saying, "Don't worry, these are normal strength. You'll be fine unless we both go mad this evening!"

We drank from the bottles, like a couple of students, and Simon continued his explanation about Juliette.

"Yes, Juliette has some interesting past. She was hired by a guy named Levi Spillmann, who later washed up on the shores of Lac Leman. They said at the time that it was a boating accident, but I'm not convinced. He was a hot shot at the labs and worked in the military systems division. I think his prior background included work in Tel-Aviv for a security systems developer, working on the development of imaging systems for military drones.

"He had figured out how to make an agricultural imaging device, which could be used to look at crop infestations. It involved a way to make a fast-moving camera think it was standing still by nestling it inside a proprietary gimbal, which helped cancel out vibrations and resulted in less blur. They put the gimbals onto light aircraft and could fly them at about 100ft and then zoom over tens of thousands of acres of farmland at 200km an hour, taking photographs with an off-the-shelf camera.

"Even at that speed, the software could spot the tiniest pests or signs of disease, and the planes can map thousands of acres in the time it takes for drones to travel just a few. Then spot and eradicate the pests. Inevitably the military were interested in it, and I think they've adapted a version for Reaper drones nowadays.

"A classic example of good science vs bad science?" I suggested.

Simon nodded, "Yes, you could be right. The good science helps stop crop wastage by spotting caterpillars.

The bad science attaches the same thing to a hunter-killer drone which can carry multiple warheads. But to the point in my story: It illustrates that Israel has become a world leader in computer vision, which is part of our quest here in Brant."

"I sense that the conspiracy part of the story is about to emerge?" I asked.

Simon nodded, "Yes. This is where things get murky. Suppose that a nation was developing the most advanced computer eyeballs, and then one of their scientists came to work for Brant. He even went on to develop and monetise one of the early products. Then suppose that another power is interested in maintaining the technological lead. A ruthless way to regain the position might be to take out the opposition."

"That's no longer just bad science," I said.

"Correct," said Simon, "It's extremely ruthless science if they start killing the scientists."

I said, "Well, Juliette didn't mention any of this to me, but she talked about Levi - she said he helped her get the job at Brant."

"That's right, but she didn't mention that they lived together. That's how she has the apartment on the Lake shore; she originally moved into it with Levi."

I recounted to Simon, "No, she did say that Spillmann had been suffering from a breakdown and that she had helped put him straight. She also mentioned that she'd broken up with someone, but not that she'd got together with Spillmann. I got the impression that she was trying to keep relationships separate from the office nowadays too."

Simon explained, "Yes, that is about the sum of it. Before she moved in with Spillmann, she'd lived here in the same block as many of us. The striking difference was that she was a native of Geneva, and so it seemed slightly weird that she'd come to these apartments rather than make her own arrangements."

Simon paused, "I can add something to the breakdown story too. Levi was a native of Israel and was conscripted into their armed services. Conscription exists in Israel for all Israeli citizens (both genders) over the age of 18 who are Jewish. The normal length of compulsory service was two years and six months for men. Levi was a smart cookie and had risen through the ranks and placed into a special secretive unit called Unit 9900."

I'd never heard of this unit.

Simon continued, "Now, the full name for Unit 9900 — the Terrain Analysis, Accurate Mapping, Visual Collection and Interpretation Agency — hints at how it created a critical mass of engineers indispensable for the future of this industry. The secretive unit has only recently allowed limited discussion of its work. But with an estimated 25,000 graduates, it has created a deep pool of talent that the tech sector has snapped up."

"But the people that work for the unit are technically soldiers?" I asked.

Simon nodded, "Yes, soldiers in Unit 9900 are assigned to strip out nuggets of intelligence from the images provided by Israel's drones and satellites — from surveilling the crowded chaotic streets of the Gaza Strip to the unending swathes of desert in Syria and the Sinai."

"Wow, it must be mind-numbing work!" I said.

Simon continued, "Correct. With so much data to pour over, Unit 9900 came up with solutions, including recruiting Israelis on the autistic spectrum for their analytical and visual skills. Unit 9900 learned to automate much of the process, teaching algorithms to spot nuances, slight variations in landscapes and how their targets moved and behaved."

"Ah, I see, so it was computer assisted? - A computer helps a human with the task?" I asked.

Simon continued, "Yes. It was taking all these photos, all this film, all this geospatial evidence and then breaking it down: how do you know what you're seeing, what's behind it, how will it impact your intelligence decisions? You're asking yourself — if you were the enemy, where would you hide? Where are the tall buildings, where's the element of surprise? Can you drive there? What will be the impact of weather on all this analysis? Computer vision was essential to this task, and teaching computers to look for variations allowed the unit to scan thousands of kilometres of background to find actionable intelligence. "

Simon paused and looked at me, "See the connection? It

was just like an extension of the work that Levi had been doing with the crop reconnaissance. Oh yes, and more than that - Unit 9900 had sold the technology to a recruitment firm. One that operates mainly in America."

But I assumed Levi wanted to get out?" I asked, "even with all of the technology success?"

"Yes. Levi was a bright guy, and I think he came up with a plan. One way for Levi to get out of the conscripted service, would be on mental health grounds, although only about 4% or people leave early this way. I'm guessing that he faked the mental stress illness to get out and to join Brant, but then need a fast-track way to prove he was fit for normal work."

"So would the hand of Brant be operating in the background somehow?"

"Almost certainly, even to the point of getting psychologist Juliette Häberli to respray him as sane and sensible."

"There's more things lurking under the surface?" I asked.

"Oh yes, you can be sure of that around Brant!" answered Simon.

Thursday with Juliette

The next day, in the Research Lab, I ran into Juliette again, this time in one of the cafe areas. She was sitting alone.

"Thank you," I said.

"Pour quoi?" she answered.

I sat down at the same table. Although we were in a works canteen, it felt more like an attractive restaurant.

"Thank you. For yesterday. It was lovely to visit the shoreline and to enjoy that evening with you." I smiled.

"Let's not get over sentimental about it," she said, "I seem to remember we were talking about Artificial Intelligence for at least half of the evening"

She was delicately nibbling chicken salad. I had a Bauernfrühstück - a farmer's breakfast - which looked altogether more heavy-duty, comprised fried potatoes, eggs, cream onions, parsley, cheese, and bacon. Despite its name, they eat it not only for breakfast but also for lunch and dinner, and I think would be typical of a one-meal-a-day diet.

"Yes, Juliette, but you shared some other things with me too. Although I hadn't realised that you were properly friends with Levi?"

"Oh, news travels fast," she said, "Yes, we become good friends. I guess someone has told you we moved in together. That's the real reason I moved from the Apartments to the lakeside. I'm sorry if I edited the story for you yesterday."

"It's entirely understandable. After all, we had only just met."

"Well, I'm going to trust you now. You seemed to be on-the-level and I need to tell someone about this. Levi told me about the Artificial Intelligence work he was doing. He was very worried that it had lost its original altruism."

"How do you mean?"

"Well, the original brief for his unit was to use AI to search for greater truths. To be life-enriching. It seemed like a good aim. Remember Google with 'do no evil?' It was that type of outlook.

"Hippy-dippy?" I asked, holding both hands out with peace signs.

Yes, until they sold the tech to a recruitment firm in

America. Levi said it had been dumbed-down - he said it was more like snake oil in the reduced form. And eminently customisable. You don't like certain last names? Reject them. You don't like certain birth cities? Reject them. You get the picture."

Juliette kept her straight face, "Levi said that the tone of the product had changed as it became more targeted toward military. It became alpha male dominated and quite strident. The main algorithms were about winning and domination.

"Winner takes it all?" I asked, grabbing a delicious forkful of the Rösti and gruyere dish before me.

"Yes. Levi used to describe the beaches of Tel-Aviv. He'd describe the hot Mediterranean sunshine and the bathers out there, with bikini-clad women carrying their assault rifles onto the beach."

"I'm sure the tourism authorities had a field day with those images!"

"They did, actually, but think of it as context. The Israeli army have very strict rules over the carrying of weapons. Taking the weapons in or out of their secure armoury involves a time-wasting amount of paperwork, waiting around, filling out forms and permission slips. On a sunny day when the beach beckons, who has the time?"

"But bikini-clad women carrying rifles onto the beach? Really?" I asked.

Juliette replied, "Weapons can't be left at home. The rules

are quite clear. If you have your weapon out of armoury, as you may well have for military exercises, then it must always remain on or about your person. You will be jailed if anything happens to your rifle, or if someone steals it while it has been left unattended."

"Now it makes sense. My friend Kyle used to work in Israel sometimes, and he said he'd always make sure that people he was talking to have the little white plastic lock securing their weapon's safety catch," I remembered.

Juliette nodded, "Yes, I visited a couple of times with Levi. I found it exhausting. It is a common sight in summer to see toned young off-duty military personnel walking about the streets, or on the beach, with their (unloaded) rifle casually strung about their person. Levi told me about a web site which showed off many such sights - it sounded sleazy, to be honest. But the people of Israel are so used to seeing military grade weapons being carried that they don't bat an eyelid."

I mused, "To the rest of the world, it serves as a constant reminder that Israel seems to be permanently at war."

Juliette continued, "I seem to spend too much time telling you brutal stories. This is another one. Levi had a girlfriend before me: Izabella Ish-Shalom, who was an officer in the Israel Defense Force. The IDF offer a very fair gender policy and so officers in the ranks comprise an equal number of 50% women, and 50% men. She was already a first lieutenant - she had one bar on her shoulders. It was because of Izabella that Levi had the breakdown."

"This was when Levi was still working for Unit 9900 in Israel?" I asked, then suddenly remembering that I'd heard it from Simon.

"Yes, although I don't think I actually told you that. I'll assume you must have got it from Simon!" she said, "Izabella had been on security duty near to the southernmost beaches of Tel-Aviv when a situation occurred."

Juliette continued, "There are no lifeguards in this area just before the rocky promontory where Jaffa begins. None of that bothered Palestinian children paddling in the shallow water. Few of them can swim. Some don't even own bathing costumes. But many are seeing the sea for the first time—enough to bring them great joy. After their parents dry them off, families may take a stroll around the central square in Jaffa. Then it is back to the landlocked West Bank."

"So, they break through the barriers to go to the seaside?" I asked.

Juliette nodded," Yes, such scenes play out several times every summer on beaches up and down Israel's Mediterranean coast. Only about 70,000 Palestinians out of the roughly three million who live in the West Bank have work permits that allow them to travel outside the territory, which is under partial Palestinian control. "

She continued, "It was sometimes on Saturdays when Israel's security forces turned a blind eye to families slipping through gaps in the security barrier. That's where Izabella's patrol was based. On the other side of the fence, friends and family members who live in Israel or unofficial tour operators wait to whisk them to the beach. The beaches are taped off and you'll even see Arab women in full Arab garb in the sea."

"I think I've seen those pictures," I could remember them from British newspapers - they used to get trotted out during heatwaves.

Juliette carried on, "Gaza has beaches, but precious little else. The territory has been under blockade by Israel and Egypt since 2007, when Hamas seized power. Israel lifted some restrictions, and some infrastructure projects were planned. But the progress quickly stalled."

I checked, "Hamas? That's the Islamic Resistance Movement, appearing after the first intifada, or Palestinian uprising, against Israel's occupation of the West Bank and Gaza Strip?"

"Correct, and it has been called a terrorist organisation by several countries, as we see the missiles and bombs flying in both directions. Now this Saturday, as was the tradition, Izabella Ish-Shalom had turned a blind eye to the Palestinians crossing into Israel. It turned out to be the weekend that a tragic bus-bombing occurred in Tel-Aviv. Many saw it as another turn of a vicious cycle. It begins with Palestinian militants attacking Israeli towns along the border with rockets or balloons carrying incendiary devices. Israel hits back with air strikes on Hamas positions. The cycle ends when Israel allows an emissary from Qatar to enter Gaza with suitcases full of cash that alleviates the suffering."

"They linked the people crossing to go bathing with the provisioning of the bus-bomb?" I asked.

"Yes, this time, Izabella got blamed for allowing the Palestinians into Israel, because they could have smuggled the bus-bomb with them. She was court-martialled and given a five-year prison sentence. It was the same length as if she had refused military service in the first place."

"Five Years? But the evidence must have been circumstantial?" I asked.

"Yes, but her crime was classed as a security crime, not a criminal one, which meant she would expect to serve the full term of five years. A criminal, on the other hand would only serve a half sentence. They put her into Ayalon prison in Ramla near Tel Aviv, which is a maximum-security jail, ruled by drugs. After being interrogated there and allocated to a single person cell, they found her hanged."

"Hanged?" I said quite shocked, "How is that even possible?"

"Levi was devastated. He said to me it was his reason for wanting to quit the military. As he had already completed over two years' service and the reason he was leaving was mental health, he could swing it to get out."

"But none of this is making sense, and even more so when we add what happened to Levi," I said.

"That's when Levi came to Brant, apparently holding it

together, but completely bitter at how Izabella had been treated."

"And hanged? Surely that is an uncommon method, especially in women?"

Juliette, added, "Believe it or not, the Swiss have written specific research about this subject. And women hanging themselves inside institutions is a particularly low percentage form of death. And when they leave no note, it only adds to the unusual nature of the event. I didn't say it to Levi, but I thought the whole situation was very suspicious."

"More like a staged death?"

"Exactly, although I've no idea why."

"Maybe the link to Levi? A warning?" I hazarded.

"That is what Levi said. Part of his paranoia was that he was afraid that someone was out to get him."

"But did he say why?"

"Only that he thought it was tied up in his research," I wondered then if he was overly stressed about work. It was difficult once we were having a relationship, if this kept re-surfacing."

I remembered that thing about not talking about your

exes too much when out dating. Then I remembered we were at work. I could also see how all of this had hardened Juliette's demeanour.

Friday with Simon

Friday evening, I was back at the Apartment and there was a knock on the door. It was Simon, with a bottle of wine.

"Hey," he said, " We can't have this. Indoors in Geneva on a Friday evening?"

"D'you know something, I'm pooped," I said.

"Fully expected after a week working here."

"I've an interesting proposition for you, you know," Simon looked with a twinkle in his eye.

"I'm not going clubbing or on the pull," I said, "I'm zapped."

"No. It's something else."

"Okay the, you'd better put some of that rather expensive white wine into one of those expensive glasses!"

"Right. A mutual acquaintance has been in contact!" said Simon, "I think you know her better than me, actually...Amanda Miller?"

I paused. I hadn't had any dealings with Amanda Miller from SI6 since my involvement with the cyber currency thing.

"Oh, Amanda? I'm not sure I'd say it was a strong link!" I said, evasively.

"Well, Amanda seems to know you pretty well. Full disclosure from me. I'm with MI5. I've been sent to Geneva to try to get to the bottom of what Brant is doing. I don't work for Amanda, but my boss does report into Jim Cavendish. I think you know him too?"

I remembered Jim Cavendish, Amanda Miller, Grace Fielding and Daniel Eversley, two people from MI5 and two from GCHQ. We'd all been together in that cyber coin situation, although I didn't expect to ever meet them again.

"Yes, I know Jim. But what on earth are you doing with Amanda and Co?" I asked.

"There's trouble at mill," replied Simon, "Or trouble at Brant, anyway. Amanda's team and an outside agency have been following it for ages. Now Brant seems to be trying to move into cyberwarfare. We think that is how Levi Spillmann got himself killed."

"But you don't know me at all. Why are you trusting me with this information?"

"I've been briefed by SI6. They've told me about your previous exploits with the cyber cash and how you helped stave off a massive economic attack by a couple of nation states. Amanda wants to see whether she can harness your considerable brain power for this situation. It's purely fortunate that you are in position here, and also that no-one would think of you as you linked to the British Government. Amanda has asked to see you tomorrow for a briefing.

"Em, tomorrow is Saturday and I'm pooped! I was planning to lay in!" I said.

"That's okay," Simon reached into his bag and produced a bar coded printout.

"Look, they have sent me these two tickets, Business class and an incredible number of air mile points, for you to fly back to London tomorrow morning, have the meeting, stay in a five-star hotel and then return on Sunday."

He showed me the tickets. And the British Airways card upgrades. I would become a gold member after just this single flight, which appeared to amass 5000 tier points.

"Er, how does this work then?" I asked.

"Take the flight and find out. You'll be in the Concorde Room at Heathrow on the way back. You'll also be working for the Crown."

And which hotel? I asked,

"You are booked into the Corinthia," replied Simon, "It's just around the corner from Whitehall."

"I'd better pack."

"The flight leaves at 09:45 and gets in at 10:30 - with the hour difference, " said Simon, "I'm arranging for us to be picked up from the Apartment at 08:00. Go hand luggage, I suggest, for speed."

Well, now I'd been in Geneva for less than a week and was already being shipped back to London for a briefing. Never a dull moment.

"More wine?" asked Simon.

Bon Voyage

"But the true voyagers are only those who leave
Just to be leaving; hearts light, like balloons,
They never turn aside from their fatality
And without knowing why they always say: "Let's go!"

— Charles Baudelaire, Les Fleurs du Mal

GVA->LHR

Saturday morning, and the airport car arrived outside of the Apartment. I'd taken a small carry-on bag, with a change of clothes, my laptop and a few relevant papers. Travelling light.

Simon seemed to have forgotten his own advice and had a wheely bag. It wasn't that big, but I reckon it would be a squeeze to get it into an overhead locker.

"Mornin'," He said, looking at my small amount of luggage, "You didn't pack a suit then?" he asked.

"No, just a change of clothes. You've got all the travel paperwork?" I asked.

"Yep, we are both good."

Then back to Geneva airport, a fast lane into the terminal, and we were drinking coffee waiting for the flight.

"Our meeting won't be at Vauxhall Cross or MI5," he explained, "Instead we are meeting at somewhere that is just a stone's throw from the Corinthia."

I was intrigued, and he announced, "Yes, we'll be meeting in The Old War Office, in Whitehall Place."

I remembered that the Old War Office was a spectacular monster of a building on the corner of Whitehall and Whitehall Place.

"I thought they had converted it to Apartments and a hotel?" I asked.

"Yes, but that doesn't stop the old site of the Secret Service from retaining a room or two!" said Simon, "We can even still get in through the old 'Spies Entrance' to avoid detection!"

We hopped the plane and were given two of the very front seats, next to one another. No middle seat, so I assumed that my BA upgrade was already working.

I asked Simon about how he'd got into the spy business, and he said he'd simply applied via an on-line form.

"It was a little bit crazy. They were running recruitment on-line. I thought it was a hoax, but it turned out to be genuine. My study had been as a research chemist, and this alternative seemed much more interesting. They specifically hired me to come along to Geneva. My mother was French, so I'm fluent, and I knew my level of research chemistry could mix the chemicals without untoward explosions, but it wasn't the clever and deep knowledge worthy of science prizes."

Simon paused to sip his airline drink, a gin and tonic.

"Yes, I came to be in Geneva, acting as a researcher, but also to delve into what Brant was doing. Unfortunately, our intelligence service thought that the original play by Brant was Big Pharma related, which is why I was embedded in that team. It turns out that the 'play' was into the robotics and cyber division - which is where you and Juliette reside."

"And how come all of the ex-pats have been pushed together into the Apartment?" I asked.

"Jury's out on that one - I guess Brant could be keeping an eye on us, like some sort of Stasi hotel, but honestly, I think it is purely economics. Get us into pleasant accommodation and we are more likely to stay. Put us with other like-minded people and we are also more likely to stay. D'you get my drift?"

"Oh yes; perhaps we're giving Brant more credit than they deserve," I added.

London

We landed in London and were soon on the Heathrow Express to Paddington. Then a taxi ride to the hotel, and even with London Prices it was considerably less than the cross-town Geneva fares.

We dropped off our bags at the hotel, and I had a fifteen-minute turn-around refresh in the plush room, which also gave me a view towards the Thames and the London Eye. Then, almost breathlessly downstairs to meet Simon and head along to the meeting.

We arrived at the Old War Office minutes later and Simon took me in through the main entrance. It was certainly a building with 'a past'. The fortunes of the Old War Office in Whitehall paralleled the trajectory of the British empire's decline and transformation.

Here we had minutes to wait, and the pace seemed to slow after a hectic day, and one in which I'd missed my planned late sleep.

We were inside this monumental building from an era that seemed almost impossibly distant, and it is now rebuilt as a symbol of a very different country. There's a huge cliff of classically carved stone illustrating the zenith of Britain's imperial pomp and arrogance.

"So what do we know about it?" I asked Simon.

"Oh, what? The building?" he answered, "Well. A decade after it was built, this colossus proved hopelessly under-scaled to cope with the bureaucracy of the first world war. Its desks would be occupied by, among others, Lord Kitchener, Winston Churchill, and Lawrence of Arabia.

"Then, later, it would also be a centre of operations during the second world war. Barely half a century after it opened, it became the backdrop for the 1956 Suez crisis and the early 1960s Profumo scandal, an explosive mix of politics, privilege, Tory hypocrisy, spying and sex."

"I could be cynical," I said, "Where once Britain siphoned wealth from the natural and human resources of its colonies, which were kept in check by a ruthless military, now it has discovered its own indigenous natural resource: prime property. The mining of Whitehall, centre of government, is one of its most remarkable manifestations."

"I agree this site is almost unbelievable. Look out of the windows of one of the suites, where moustachioed warlords sat around maps with pots of tea, and glimpse red and silver from the Horse Guards, mounted on glossy black beasts, their breastplates gleaming, their cutlasses glinting in the sun. They stand sentry at the gates to what were once Henry VIII's tilting grounds, the fields on which they held royal tournaments."

"So, I take it the apartments are selling well?" I asked.

"Flying from the estate agents! Imagine how much cleansed money can be deposited through purchase of one of these properties?" said Simon, "And next door is Inigo Jones's 17th-century Banqueting House, with its ceiling painted by Rubens. Just outside, Charles I was executed for treason. To the right is Trafalgar Square, to the left the Palace of Westminster — and Downing Street is a couple of hundred yards away."

"I can't believe they would sell it off?" I asked.

"Truth be told, neither can I. This building is at the UK's seat of power, the nexus of government, military and establishment. I can't imagine that any other government anywhere in the world would be prepared to flog off a site like this and yet, here it is, a building sold on a 250-year lease for £350m, enough to buy about a tenth of a new aircraft carrier."

"Dodgy dealings?" I asked,

"Who knows? Crazily, the sale was part of a £5bn Whitehall downsizing aimed at reducing running costs. Civil servants who worked here decamped to other locations. The building has an undeniable presence and was sold by the Ministry of Defence to a Spanish/Indian consortium."

"There's a hint of Trumpism in these property moves?" I

suggested.

Simon smiled, "This has been going on for ages. Britain has the super-prime properties and now if one is rich enough, it's possible to cosy up to the Prime Minister's residence, and the River Thames in arguably a better location. Look: The once fashionable 'new Rome' of an Edwardian baroque style has been emphasised in this rebuild. It has an undeniable presence, an Edwardian sense of solidity, ambition, and self-importance. At the heart of the building is a grand staircase in richly marbled Renaissance style, a design that exemplifies the Edwardian civic style, splendour, and grandeur at the expense of taste or subtlety, but undeniably impressive."

"So, the target market is brash?" I asked.

Simon replied, "I'm not sure I'd say 'brash'. Maybe 'unsubtle'. Along the miles of corridors are offices once occupied by Kitchener, Lloyd George, TE Lawrence, Churchill and Profumo. John Profumo took his 19-year-old lover Christine Keeler to the grand Haldane Suite, the imposing office of the secretary of state for war, named after its first inhabitant, the 'philosopher-politician' Lord Haldane. Keeler was also sleeping with Russian naval attaché Yevgeny Ivanov, potentially threatening national security. Profumo resigned in 1963 when it became clear he had lied to the House of Commons about Keeler."

A woman's voice chipped in, "Charles de Gaulle commented archly at the time: 'That will teach the English to try to behave like Frenchmen.'"

We both looked around and I recognised Amanda Miller.

"Hey Amanda, it's great to see you."

I wasn't sure whether to *'bise'* her on the cheek, hug her or simply shake hands. I could see she sensed the confusion, and she came in for a hug.

"A week in Switzerland and your whole etiquette system has gone crazy!" she added.

She shook Simon by the hand, and I sensed they knew each other less than I knew Amanda.

"Well, it's an honour to be greeted by you, in person!" I said, "I trust that life is good?"

"Yes, everything is fine. You'll be surprised to see that I've been getting the band back together for this Brant challenge."

She led us to a lift and pressed the up button. "Top floor," she said.

"I knew it!" I smiled, my excitement rising.

Amanda's briefing

We walked along the corridor and I spotted that we were going into a room at the end. Anyone who knows hotel corridors will know that Rooms At The End are usually Bigger and Better.

I wasn't expecting the sight inside. There sat Grace Fielding as well as several others I didn't recognise.

"Wow!" I grinned, "This is amazing. It's also slightly worrying. It's great to see you all! Amanda certainly knows how to pull strings and influence people!"

"And you must be Simon Gray?" asked Grace, looking over at Simon, "We'll all introduce ourselves."

They went around the table, and they introduced me to several new people, who seemed to be important to Amanda.

"Yes, hello, My name is Clare Crafts and I've worked with Amanda on and off for several years. Several of us here work for an outside company called The Triangle and occasionally pitch in to help with something unusual that Amanda or Grace want."

"I'm also from The Triangle, my name's Jake Lambers, and I've a background in journalism."

Clare added, "Oh yes, I should have said, my background is marketing and computer graphics."

"And I'm Dave Carter - call me Bigsy. Another member of the Triangle and a bit of a whizz with computers and suchlike."

"And my name is Christina Nott, best described as a world traveller and quite linked in with the Triangle and their exploits."

An American spoke next, tanned, softly spoken and sounding as if he could be from Texas, "My name is Chuck, Chuck Manners and I'm from the other side of the pond, and I think Amanda has called me in because I've worked with missile systems."

Amanda smiled, "I've already briefed the gang here about you two, we gave you both a rather fine profile."

Simon looked over at me. "I'm not sure that we'll remember all of your names!" he said.

Clare pushed a piece of paper across the table, "Here, I've written us all down on this." She pointed to a neat map of the table with everyone shown in the correct position with their names.

Amanda spoke next.

"Thank you all for coming along to this briefing. We have been tracking the exploits of Brant for several years, and indeed, the exploits of their owning company, called Raven Corp."

Amanda continued, "We think Brant is building a new weapons system. They have based the development in Switzerland, which is out of the European Union's jurisdiction and outside of NATO. Swiss cooperation with NATO is based on a longstanding policy of military neutrality and areas of practical cooperation that match joint objectives.

Jake interrupted, "But I seem to remember that Switzerland has supported NATO-led operations in the Balkans, as part of the Kosovo Force. And Switzerland also supported the operation in Afghanistan up to 2007, I think?"

Amanda continued, "Yes, Switzerland shares its expertise with NATO by offering education and training to Allies and other partner countries. Some areas are humanitarian missions, international humanitarian law, human rights and civil-military cooperation, search and rescue training, security policy, arms control and disarmament, and transparency and democratic control of armed forces."

Amanda said, "But let's focus in on what Brant has been working on, and the area that Matt finds himself in the middle of. It is similar work to that conducted by Levi Spillmann - You'll remember it from my earlier briefing.

Sorry, Matt and Simon; you missed it."

Amanda continued, "This was the image recognition system that could work from a plane to provide bug identification on crops. The extension that Brant wanted was to make the system able to perform facial recognition. More than that, they wanted it to have AI alongside the Facial Recognition."

"That's like the Chinese system that they use for reviewing people in crowds?" asked Bigsy.

Amanda nodded, "Yes, that's right, Bigsy, it's made by a company with the unfortunate name of Hikvision, which grew out of a Chinese government research institute, but is now only partially state-owned. They have a growing international presence, supplying video surveillance infrastructure from Brazil to South Africa and Italy. It is one of a group of Chinese tech companies — along with Dahua, SenseTime and Megvii, who are among the world's leading artificial-intelligence software vendors.

"Yes, and they've been blacklisted by Washington over alleged involvement in supplying technology used in Xinjiang, you know, where they operate the detention camps," added Chuck.

Grace added, "Facial recognition is also being used to extend surveillance in new ways, such as tracking the classroom behaviour of students. China's rapidly expanding network of facial-recognition cameras means many people are now subject to mass identification. China's police force already holds the world's biggest national database of over one-billion faces, captured for the national ID card system. Depending on the quality of these photos, they can be matched to people posing for a face scan.

Amanda added, "The industry is paying attention to the increasing complaints. SenseTime - the world-leading AI company based out of Singapore announced it will lead a consortium of 27 companies to help set national standards for facial recognition under a government standards body. In China, the state media reported that the country had over 20-million AI-enabled cameras.

"While debate over the use of facial recognition in the EU and the US focuses on the privacy threat of governments or companies identifying and tracking people, the debate in China frames around the threat of leaks to third parties, rather than abuses by the operators themselves."

"So what, exactly are Brant trying to do?" asked Jake.

Amanda said, "Two things. One, to sell their technology to the highest bidder and Two, to boost their own share price. Brant - which is a Raven Corp company makes no bones about these basic objectives."

"Now this is where the sadly departed Levi Spillmann's information comes in. He reckoned that the technology was intermittently flawed and that it could not be usefully interfaced to AI systems. He wrote a paper about it while he was still working for the Israeli Defence Force. We couldn't get the paper, but our colleague Christina Nott here has some special friends and they managed to track down a copy,"

Christina smiled, "Yes, the Russian Intelligence had some embedded operatives in Unit 9900, and they found out that there were several flaws in the AI being used. They were derived from Levi's paper"

Christina said, "Let me list them for you." She stood and walked towards the smart board in the room. She reminded me of a cat in the way she moved. She didn't so much walk as slink her way to the board. She was also wearing black leather trousers which hugged her form. I noticed Simon's intake of breath.

Christina did something with her phone, and a projection of a slide appeared.

- Male Gaze
- Racial Profiling
- Ruthless
- Fallible

Christina began, "Let's talk about each of them in turn."

"First: Boys will be boys. The AI is heavily slanted towards the male gaze. What do I mean by that? Well the AI will attempt to recreate a whole human form from the data it is provided. Give it a man's head and it will generate the form of a man, typically wearing a suit and with a tie. It uses a huge amount of data to provide this, and it is derived from many images of people stored and then re-processed.

"But give it a woman's head and it will generate a woman in a bikini. Some of you might have heard about this from a leak about a well-known US female Senator. Her head was provided and a similar AI system generated a likeness of her in a bikini."

"I see, so we are dealing with male bias in the AI Algorithms?" asked Bigsy.

"Yes, in the algorithms and in the datasets being examined," answered Grace.

Christina continued, "Second: Racial profiling. In the way that Levi's camera swooped over fields looking for bugs, it needed to edit out some benevolent species. It would ignore bugs considered good and filter for just the 'bad bugs'."

"I see," said Bigsy, "So the system can be tilted?"

"Correct," said Christina, "Imagine if the human scanning algorithms had been educated about certain racial types. The 'good' ones could be ignored, and the 'bad' ones selected. That's a form of racial profiling and highly undesirable."

"The next point is the ruthlessness of the algorithms. They are never 'set to stun'. They make binary decisions. Good or Bad, where Bad leads to something unpleasant."

There was silence in the room at this statement.

"The last point is that the algorithms are fallible. For speedy inspection of crops, it doesn't matter if the accuracy is somewhere in the 80 to 85% range. It is still extremely useful. But for humans, an 80% accuracy rate is unacceptable. It's like a 1 in 5 failure to correctly assign."

"So, what happened to Levi's paper, then?" Asked Jake.

"Nothing," answered Amanda, "It was buried - we think it was because there was a deal being done with an American firm. Our supposition is that Levi was threatened because he was exposing the weaknesses in the system, and then his girlfriend Izabella Ish-Shalom was framed, jailed and ultimately killed. Someone was leaning on Levi."

"Going after family and friends is a particularly vindictive form of crime," said Grace, "I've profiled people and even the Mafia won't normally resort to such tactics."

"It suggests Russian inspired violence," said Amanda, looking towards Christina, "They are more likely to go after family members to threaten someone."

" I don't think it is," said Christina, "It's too discoverable. Russia would use poisons rather than movie-style deserted boats on the lake. I think it will be a Sicario - a drug cartel hit-man. They don't mind spraying bullets around and leaving messy trails. Russia is all about deniability nowadays."

Chuck interrupted, "Yes, I think Christina has a good point; the soldiers along the Rio Grande are just as messy with their actions. Most of the time they operate without context. Do something to someone but no reasons are ever provided."

"If that is true, it implies cartels are interested in the AI-tech. But why?" asked Christina, "They are not exactly waging wholesale battleground warfare."

"I agree," said Amanda, "Which is why we need to do some more digging."

"Here's my requests," She nodded towards Grace, who flipped up a new slide, across the top of the one that Christina had projected.

SECRET DATABASES
- Secret Databases inside Brant
- Databases used for surveillance and to augment the AI
- Simon Gray has no access to Brant secure levels, nor will he get them
- Juliette Häberli has direct access to the Brant secure levels
- Matt Nicholson can expect access after 6 months probation.

BRANT SECURITY
- Strong but potentially capable of being broken into from inside.
- Security expertise from Daniel Rayleigh?

MISSILE TECHNOLOGY
- Presumed to be missile-mounted, but increasingly likely to be drone or UAV mounted
- Use Chuck Manners expertise with missiles to identify likely plans

AI AND FR TECHNOLOGY
- Explore and validate Levi's thinking
- Matt Nicholson to investigate
- Decide how to deter development

POTENTIAL CLIENTS FOR THE WEAPONS
- Recognised superpower (US, Russia, China)
- Unstable state (e.g. North Korea, Pakistan)
- Sanctioned State (e.g. Saudi Arabia, Iraq, Afghanistan)
- Terror endeavour (e.g. drugs cartel, traffickers)

BRANT MOTIVATION
- Political gains
- Financial gains through share price
- Financial gains through sales.

"There," said Amanda, "We can have a clear look at the options and who needs to be involved."

Simon and I looked at one another. There must have been work on this before the meeting, no one could predict all of that in real time.

"No disrespect to Daniel, but I think you could use my friend Kyle again," I said, "Kyle is a security guru and has also worked in Israel and knows many of the systems. I think Kyle is working in Cambridge now."

Grace said, "That is a good idea. Kyle was involved with that Coin setup and was very helpful as we worked out ways to break past the security systems."

Amanda replied, "We wanted some untraceable people who could think outside the box. I remember Kyle is just one of those people. I think he and Bigsy will get on well. Can you get into contact with Kyle, Matt?"

"Sure, I have his number in my phone. But how long will you need him?"

"I'd say a week on site - in Geneva, and then we could operate the rest remotely," said Amanda.

"Okay, I don't know how he will be able to work that," I answered, "Let me call him."

I walked out of the main meeting room into the long corridor. I located his name and dialled him.

He answered, "Hey Matt, it's been a while!"

"Yes, it has, I thought I'd see how you were doing and whether you were up for some foreign travel?"

"Ha, I've just got back from Saudi. Strangely enough, I was helping on an American project. Kinda creepy. Where would you like me to go?"

"Geneva. I'm working there now and we have a knotty problem. It involves some people who were involved with our cyber coin clean up."

"Hmm, that sounds very interesting. Is Amanda involved?"

"Yes, she is as a matter of fact."

"Okay, I'm in. It must be something good. I've just made a shedload of money from the Americans and really I need some downtime. I guess seeing a buddy and helping Amanda could be relaxing. Geneva's a pleasant city too. That lake, those mountains. Toblerone."

"Okay, Can I tell Amanda that you are in? A week then remote?"

"Sure thing. I can get a new Swiss knife whilst I'm there. One with all the attachments."

I returned to the room, where things had broken down into several independent conversations. I told Amanda, and she looked pleased.

"Okay, everyone, now we have Kyle Adler plus the inimitable creativity of Bigsy."

"Then a missile expert in the form of Chuck Manners, and Brant insiders in the shapes of Matt and Simon."

"Add the analytical skills of Grace Fielding, Jake Lambers and Clare Crafts, and we have an excellent team."

"Er, what about Christina?" asked Clare.

"Christina will probably prefer to operate in the shadows, I suspect," answered Amanda.

Christina's smile didn't give much away.

Corinthian

Sunday morning and it didn't feel as if I'd had a weekend. The trip to London felt like a business trip.

Then Bigsy had suggested we all go to the bar at a nearby hotel - Guess which one? The Corinthian. Of course, the group from the Triangle were all used to one another and there was some good banter. Chuck Manners left with Amanda Miller and I have a feeling that they could be an item.

By the time I got to bed it was around three am, and I wasn't even the last to leave, with Christina, Clare, Bigsy, Jake and Simon still managing to keep things going.

Of course, I couldn't sleep in on Sunday. I woke up and then couldn't get back to sleep. I looked at the clock and it was 08:30, which would be 09:30 in Geneva. I had the rest of the day because we were catching a late flight back. Simon had arranged for our rooms to be booked through until 16:00, and then for us to be taken back to the airport direct by a car. I had all of London before me, but like a saddo stayed in my room, flopped on the bed, with room service and a couple of movies.

Simon didn't ring, and I assumed he'd decided to leave me in peace. I think London was more of a novelty for Simon compared with me, as I'd lived right in the centre for several years. The whole of the Triangle team were also Londoners, with apartments and an office right by the River.

Yes, I would be the sad one, staying in my room to decompress after an eventful week. I eventually packed my small bag and headed down to the lobby at around 15:30, to wait for Simon.

To my surprise, he appeared with Clare, from the elevators.

"Clare offered to show me around some of London today," he explained, "We took a riverboat to Hays Galleria, she showed me the Triangle Offices and then we walked along the South Bank right through Tower Bridge and up towards Parliament. Clare tells me she has worked there!"

"Yes, then we could grab a tube back from Westminster to Embankment and back to the hotel," added Clare, " I think we maximised the time quite well! Oh, and we ran into Bigsy in the office, even on a Sunday!"

I impressed with how much they'd done and they both looked quite refreshed.

"Drinks?" I asked, more out of politeness.

"No, I'm good," said Clare.

"Me too," answered Simon, "I've brought my bag down and I'm ready to hit the trail. The car should be here in a few minutes."

"Sunday more people will come back to London - you should have a pretty good run to the airport," said Clare, "What is it? T5?"

We both nodded, Simon spoke, "Yep, the hotel arranged the car, I think it's an Über."

A concierge approached us, "Mr Gray?" your car is here.

We said our goodbyes and as we did so, Clare thrust a small piece of paper into my hand. I placed it in my pocket. We were soon on the way.

"Clare seems nice?" I said.

"Yes, she stayed over...it was more convenient for our plans and everything."

"Almost too much information," I said.

Eight pointed star

Clare's hurriedly produced paper note had asked me to dial into a call with The Triangle at 7am British time. 08:00 Geneva. That was seriously early. Her note said I wasn't to tell anyone, and I was to use a headphone.

Sure enough, I dialled in. Clare and Bigsy were waiting for me.

"Yes, Matt - Amanda and Grace only have good things to say about you! And sorry if I was slightly furtive yesterday," said Clare.

Bigsy replied, "It worked out well, with Clare and Simon visiting the offices. Simon now has a little extra App on his phone. We can track him and intercept his calls."

"He seems to be who he says he is. Although he was packing quite a lot of extra equipment, beyond the normal clothes, phone and laptop. An SLR, a drone, and a small package of electronics that looked like a key copier."

"But if he works for SI6, wouldn't we expect that?"

Clare spoke, "Maybe, but the stuff looked like the stuff Bigsy uses, quite high end. It didn't look like 'government issue', that's for sure."

I asked, "Any other obvious signs from him?"

Clare replied, "No, he seemed to be on the level. We will cross-check back to Amanda about him. Simon gave his back story, and it was that he'd applied to join MI5 through an on-line form."

She added, "It's possible, but it's a far less likely route than MI5 pretends with its marketing. More often successful candidates win because of 'wheels within wheels.' "

I suddenly realised that Clare had been running identification on Simon Gray. I couldn't understand why, given that he was one of Amanda's trusted agents.

Bigsy spoke, "Amanda asked us to check out Simon. She said she already knows and can vouch for you, but that Simon has somehow 'landed' as a conveniently placed agent in Geneva. She is not sure that everything adds up."

"Well, he seems nice as pie and a straightforward lad," I said.

"That's what we all thought initially; it was Christina who spotted something unusual. During the evening - I think it was after you'd gone to bed. He was leaning over towards her and she could see inside the neck of his shirt."

"Em, isn't it usually the other way around?"

"Now now! Christina could see a small tattoo. It was near his shoulder and she said it was unmistakably the tattoo of someone high-ranking who had been an inmate of the Russian prison system."

"What was it? A number or something?"

"No, a small star, with eight points. It's a very typical tattoo, but not one you'd get in British jails."

"But why would Simon Gray have that tattoo? And isn't it a giveaway? Not very spy-like."

"I agree but think about it. Most of the time it's covered up. When it is visible, you'd have to know a lot to decode it. He could explain that he was interested in astronomy, or something like that."

"But is Christina certain? I mean if she only saw it for a split second?"

"Ahem, I think I had more of an opportunity when he came back to stay in my hotel room," explained Clare, "I can confirm that there were two stars, one on the front of each shoulder."

"Moving swiftly along," said Bigsy, "It just raises some suspicion with us. Is Simon what he says, or is there a different explanation? Anyway, I've sent you a small gift. It'll arrive with Amazon in the next day or so."

"What is it?"

"A bug detector, for you to scan your room. Maybe also scan the office."

"But will I know how to use it?" I asked.

"You will, I've selected an inexpensive model that will detect anything they are likely to use. It will flash and bleep at you."

"They?" I asked, "who are they?"

"We'll assume it is Brant at the moment.," answered Bigsy.

Clare said, "You know something, try to check Simon's flat as well."

"This is getting more complicated, " I said.

"Well, it should make you feel more like a secret agent!" answered Bigsy.

Selexor and Createl

Monday, and I caught the bus to work again, with Simon. I mumbled that I'd failed to get my lay in and that I'd been woken early with a phone call. We chatted briefly about the weekend, but he said we should probably keep quiet about it in the office.

When I reached the Research Lab, Juliette, Herman Schmidt, and Rolf von Westendorf were chatting to Amy van der Leiden. I realised that Monday mornings were the time for a team meeting. I'd missed it last week because I was the new boy, but now I'd need to pay attention.

Schmiddi spoke first, " My god, you don't look as if you've just come back from a relaxing weekend, more like you've been out for an evening at the Mandarin Oriental!"

I laughed and thought I'd need to amp it up a notch so that they wouldn't think of me as a part timer.

"Au contraire," I said, "I've just had a relaxing weekend.

So nice to enjoy the splendour of Geneva!" I felt smug with myself that I'd even managed to build a couple of French words into my riposte.

Amy had news from the big boss. They had all switched into speaking English now, to keep me on-side.

"Kjeld asked to see me on Friday evening," Amy explained, "He wanted me to re-visit the commercialisation of the work conducted by Levi Spillman. Brant's lawyers wangled it that Brant would hold the patents for all the work that Levi had conducted in the prior three years. It included the intellectual property rights for the bug-scanning algorithm, which was then linked into Selexor."

"This next piece is confidential," Amy explained, "Raven are planning a takeover of Selexor. It means Raven/Brant would own the whole of the production chain. 'Soup to nuts', I think the Americans call it."

Amy continued, "The original licensing to Selexor of Levi Spillmann's Createl AI system was something of a coup for Brant.

"I remember- it was a big deal when Levi was hired, but it seems to be all about the product, rather than Levi himself," said Schmiddi, then adding, "Oh- I'm sorry Juliette - I don't mean anything by that."

Those around the table looked awkwardly at Juliette. It was obvious that everyone knew that Levi and Juliette had once been together.

"Look," said Juliette, " I'm over Levi now. I knew he was a damaged person when we first got together. To be frank, I should probably have maintained a professional distance. Let's clear the air about all of this so that we can move on objectively."

It blew my mind how Juliette could talk so objectively about an ex-lover. I had to regard her as putting on a front to protect herself.

Amy continued, "Selexor used Createl artificial intelligence in its hiring system. Selexor has become a powerful gatekeeper for some of America's most prominent employers, reshaping how companies assess their workforce — and how prospective employees prove their worth."

"Yes, many top firms use video interviews run by Selexor nowadays as an initial filter," agreed Rolf.

He added, "Selexor's system uses candidates' computer or cell phone cameras to analyse their facial movements, word choice and speaking voice before ranking them against other applicants based on an automatically generated "employability" score."

I was thinking this sounded too good to be true. I remembered the old phrase about deals: 'If it looks too good to be true, it probably is too good to be true.'

Amy added, "Selexor's AI-driven assessments use Createl and have become so pervasive in some industries, including hospitality and finance, that universities make special efforts to train students on how to look and speak for best results. More than 100 employers now use the system, including some of the biggest names and more than a million job seekers have been analysed."

I cut in, "And that must mean they are building up a huge database of individual profiles too; the whole system feeds itself and increases in value as more names are added."

Juliette commented, "Yes, that was what Levi was aware of. He knew that his original bug-hunting algorithm had been hacked to search through faces and that the Facial Recognition was extremely rudimentary. He used to say it was about as good as that used in a modern camera, great to spot the left eye for focusing, but not much more."

Amy nodded, "Precisely, and that is why some AI researchers argue the system is digital snake oil — an unfounded blend of superficial measurements and arbitrary number-crunching that is not rooted in scientific fact. Analysing a human being like this, they argue, could end up penalising non-native speakers, visibly nervous interviewees or anyone else who doesn't fit the model for 'look' and 'speech'.

Juliette spoke, "In addition, the so-called customisation of the system means arbitrary prejudices can be built into the selection. Imagine if it said 'no-blondes' or no-one with a high body-mass-index. It's like those old-fashioned high street recruitment agencies who could pre-filter for the prejudices of their clients."

Amy continued, "Selexor, they argue, will assume a critical role in helping decide a person's career. But some experts doubt it even knows what it's looking for: Just what does the perfect employee look and sound like, anyway?"

Schmiddi spoke, "I found some research quotes - Look, here is an example: 'It's a profoundly disturbing development that we have proprietary technology that claims to differentiate between a productive worker and a worker who isn't fit, based on their facial movements, their tone of voice, their mannerisms,' said Tiffany Shields, a co-founder of the UltraCognos Institute, a research centre in Stamford, Connecticut."

"It's pseudoscience. A license to discriminate," Juliette added. "And the people whose lives and opportunities are literally being shaped by these systems don't have any chance to weigh in. And that's where Levi was angry. He wasn't about to throw in the towel. He wanted to show everyone that the connecting together of Selexor with Createl was a huge system kludge."

"But I guess we have to roll this forward?" I asked, "If the Createl technology is to be used in Human-Computer Interfaces? The design would be deeply flawed?"

Amy cut in, "Yes; that is the difficulty. Kjeld Nikolajsen wants us to push to get the Human to Computer link designed and is not prepared to listen to arguments about parts of the system being defective. He thinks that any kinks can be ironed out later in the design and marketing process."

Schmiddi spoke again, "I was looking at the reports from Selexor. They seem to have a good PR firm. Harry Stensen, Selexor's chief technology officer, said that the criticism of Createl is uninformed and that most AI researchers have a limited understanding of the psychology behind how workers think and behave."

Schmiddi continued, "Stensen compared the Createl algorithms' ability to boost hiring outcomes with medicine's improvement of health outcomes and said the science backed him up. The system, he argued, is still more objective than the flawed metrics used by human recruiters, whose thinking he called the 'ultimate black box.'"

We all watched Schmiddi reading his report on his laptop, " 'People are rejected all the time based on how they look, their shoes, how they tucked in their shirts and how 'hot' they are,' Stensen told The Washington Post. 'Algorithms eliminate most of that in a way that hasn't been possible before. The AI doesn't explain its decisions or give candidates their assessment scores, which Stensen called 'not relevant.'

Rolf asked, "Wasn't Stensen the man who said, 'When 1,000 people apply for one job, 999 people are going to get rejected, whether a company uses AI or not.' ?"

Juliette added, "Yes, in the literature for my psychology work I see increasingly regular quotes that these inscrutable algorithms have forced job seekers to confront a new interview anxiety.

"An example, from right here at Brant: Lilja Jussila, a

University of Connecticut senior studying math and economics said she researched Selexor and did her best to dazzle the job-interview machine. Lilja answered confidently and in the time allotted. She used positive keywords. She smiled, often and wide."

Juliette continued, "But when she didn't get the job, she couldn't see how the computer had rated her or ask how she could improve, and she agonised over what she had missed. Had she not looked friendly enough? Did she talk too loudly? What did the AI hiring system believe she had gotten wrong?"

"So what were the reasons?" I asked.

Juliette remembered, "Lilja said that maybe one of the reasons she didn't get it was that she spoke a little too naturally. She didn't use enough big, fancy words. It's like not 'playing the game'."

"I remember the case," said Amy, "It made it to the top because we were really desperate to hire someone with Lilja's skill-set. My theory was that it was simply her international Finnish-sounding name that threw her out."

Schmiddi added, "Selexor said its system dissects the tiniest details of candidates' responses — their facial expressions, their eye contact and perceived "enthusiasm" — and compiles reports companies can use in deciding whom to hire or disregard."

He was still reading from the laptop report, "Job candidates aren't told their score or what little things they got wrong, and they can't ask the machine what they could do better. It's claimed that it would be the first stage of gaming the system. Human hiring managers can use other factors, beyond the Selexor score, to decide which candidates pass the first-round test."

Rolf added, "The advertising says that Selexor employs superhuman precision and impartiality to zero in on an ideal employee, picking up on tell-tale clues a recruiter might miss. Here:"

He handed out a flyer advertising Selexor. It was a photocopy of a page from Winners magazine:

Selexor

Selexor's prospects have cemented it as the leading player in the brave new world of semi-automated corporate recruiting. It can save employers a fortune on in-person interviews and quickly cull applicants deemed below the standard. Selexor says it also allows companies to see candidates from an expanded hiring pool: Anyone with a phone and Internet connection can apply.

Armanis Winterhall, Selexor's chief industrial-organisational psychologist, told Winners magazine the standard 30-minute Selexor assessment includes half a dozen questions but can yield up to 500,000 data points, all of which become ingredients in the person's calculated score.

The employer decides the written questions, which Selexor's system then shows the candidate while recording and analysing their responses. The AI assesses how a person's face moves to determine, for instance, how excited someone seems about a certain work task or how they would behave around angry customers.

Those 'Raw Response Units,' Winterhall said, can make up around a third of a person's score; the words they say and the 'audio features' of their voice, like their tone, make up the rest.

'Humans are inconsistent by nature. They inject their subjectivity into the evaluations,' Winterhall said. 'But AI can data analyse what the human processes in an interview, without bias. And humans are now believing in machine decisions over human feedback.'

Riding with Juliette

Monday evening, I was heading for the office exit, when I again ran into Juliette.

"Hi, I was hoping to see you again," she said.

"Hey." I put my thumb out like a hitchhiker.

"Certainement," she said, and we walked together to her Porsche.

"People will be talking," I said.

"Non, they saw me with Levi enough, but it was never an issue. The Swiss have certain French attitudes,"

I jumped into her car and then did something rather strange. I waggled the gearstick.

"Oh, I'm so sorry!" I said, "It's just that I'm in the UK driver seat and I'm so used to - you know - checking."

She laughed. "I know - I've done that when I've been in

England!" she admitted, "It does look rude though doesn't it? Like you don't trust the driver!"

She indicated and pulled away, on the road back into Geneva.

"You know how to make me feel bad!" I said. It pleased me that Juliette was less reserved toward me now.

"Yes, I was hoping to have a further chat," she said, "About some things we found out about today."

"Okay," I said, "Where shall we go?"

"How about yours?" she asked.

"Er, maybe we could go to a cafe or something, " I suggested. I was worried after Bigsy had told me about possible listening devices.

"I'd love to invite you back, just not today, though."

She looked at me, amused," It's because of what I said to you the other day at La Terrasse? About you falling for me? And the rebound girlfriend and all of that?"

"No," I said, "Honestly, it's not. I just have some things to do before I can invite anyone around to the apartment."

"Oh," she breathed out like a long sigh, "It's just that I was revising my viewpoint of you somewhat."

She was indicating to pull in to a roadside cafe.

"Here," she said, "It's my little refuge place. I live just along the road from here and visit this place all the time."

We left the car, and I could see a small lakeside cafe, with another view towards Jet d'Eau - this time the fountain had Geneva in the background. I tried to figure out where we'd visited when we'd gone to the Terrace place on the other bank.

"La Terrace is over there!" Juliette pointed and I could suddenly see the row of little yachts which I remembered next to the other restaurant.

"We'll sit outside?" asked Juliette, I nodded my agreement, and we walked to a wooden table which had a sunshade advertising ice cream.

"Here?" suggested Juliette, "We won't get as much sun as on the other side, but it is still pleasant here."

"So you've intrigued me," I said, wondering what Juliette wanted to see me about.

"This is difficult," she said, "I must trust you now, and that you can be discreet. Also that you will not think of me as a crazy person when I tell you this."

I held up my hand like in a courtroom, "However it sounds, you have my trust," I said.

"Okay, let me begin. This will sound strange, but I've wondered about some aspects of what happened to Levi."

"You told me that last week,"

"Yes, but there is always more, and I haven't felt I could share it with anyone from the Lab until now."

"What is it?" I asked; I could see she was still nervous to tell me.

"I think that Levi's death was engineered," she said, "Like it was staged. He was so good with boats, and I can't imagine he'd do anything stupid when he was out on the Lake. I've been out with him and he was so confident around sailing boats and motorboats. It makes little sense."

"For example, when his body was discovered, he wasn't wearing a lifejacket, although he always wore one on the yacht when I was with him. He used to tell me I had to wear one too. As a matter of fact, he was quite proud of his Helly Hansen jacket, which had some gizmo to make it inflate automatically when it was in the water. It was also bright orange which he said was good for visibility in the water."

"So he was very safety conscious?" I asked, aware we were having a long conversation about her ex. Usually a no-no, especially as she'd also said she quite liked me.

The woman from the cafe had come out. She greeted Juliette like a family member and then asked us both - in English - what we'd like. Juliette ordered coffee and a cake, and I asked for the same.

"Well, he was very proud of his yacht and liked to take

people out on it around Lac Léman. I was impressed, because after a short while you can be so far from shore that even the places start to blur into one another. It's much more like being at sea, than just on the Lake. Remember its 73 kilometres long and around 14 kilometres wide."

"I know, when you fly into Geneva, the whole lake is visible from the air and you get a sense of its size."

"Well, someone who used to like going yachting with Levi was Simon Gray. He was also a yachtsman, and he certainly seems to know his way around a boat. Simon explained he'd been crew on a Barents Sea yacht."

"Barents Sea?" that's off to the east of Norway, isn't it? And almost in the Arctic Circle?"

"Yes. It's not an obvious place to go - and yet Simon insisted he was on a super yacht there. He said he'd been crewing whilst he was doing his gap year."

"Most people crew in the Med?" I asked, "Why would anyone do this?"

"I know. I thought it was unusual. But then, on another day, I heard Simon on the phone. I know he speaks English and French, but I was surprised to hear him speaking - I think it was - Russian."

"That is interesting; I'd never have detected that Simon wasn't British. He has such a strong slightly 'posh boy' accent - and carries himself like he's been in the Army as an officer, maybe."

"Well, it got me thinking. Levi and Simon went out together several times on Levi's yacht. I racked my brain to remember if he'd said anything about Simon on the date he disappeared. I'm slightly ashamed to say I couldn't remember and even when I looked through Levi's diary and phone records, I couldn't find any reference to it."

"So, are you suspicious then?" I asked, all too aware of what Christina and Clare had said.

"I am, but I don't know on what grounds. Simon seems perfectly pleasant if, nowadays, a little cautious around me. I've always put that down to him being aware of my friendship with Levi. Oh, and the way some men are around me when they like me but don't want to say anything."

I realised Juliette must keep her self-protection at a fairly high level.

"Juliette, do you mind me asking you something personal? You might find it impertinent?"

"No, go ahead and ask," she said.

"Have you always been this way? I mean about men. You seem very guarded? Or is it since Levi's death?"

"I've tried to answer that myself. My professional life means that psychiatric counselling of male patients means they turn me into one of their friends. Some of them objectify me too. Those are valid reasons that I've been careful. But since Levi's death, I think I've been extra cautious. I mean, I gave you 'the speech' when we were at La Terrasse, and yet we'd only just met."

"I respected you for your forthrightness. Like now, actually."

"So is there anything I can do to help?" I asked.

"Well, you've already listened to me, that's good. But I wonder if you could also check for any unusual behaviour from Simon. Maybe you could ask him something about boats and see how he answers. I suspect his motives for being in Geneva, and I even wonder whether he has been set up to monitor the other inhabitants of the apartments."

"Okay, I'll see what I can do," I said. We'd both finished our cake and coffee.

"Well, if you won't show me your apartment, at least I can show you mine," Juliette said softly.

World ain't real, (I think)

Stop, what the hell are you talking about? Ha
Get my pretty name out of your mouth
We are not the same with or without
Don't talk 'bout me like how you might know how I feel
Top of the world, but your world isn't real
Your world's an ideal

So go have fun
I really couldn't care less
And you can give 'em my best, but just know

I'm not your friend
Or anything, damn
You think that you're the man
I think, therefore, I am
I'm not your friend
Or anything, damn
You think that you're the man
I think, therefore, I am

Billie Eilish

Box ripping

Suffice to say I didn't get back to the Apartment until early Tuesday morning. I called a taxi from Juliette's and was driven back.

Outside my door was a small package from Amazon. The brown box with its cardboard smile was facing me as I turned the corner and I immediately realised it must be the gadget from Bigsy. A few minutes of box ripping, and I had the little unit which reminded me of an old-fashioned voltmeter. I put in the batteries and switched it on, expecting it to be unspectacular. How wrong I was!

The unit immediately started making a bleeping sound and a row of lights lit up on the front. I almost dropped it. I looked around for the off button and pressed. Now to find the tiny manual, which was written in five languages.

It explained that I could locate sources of radio frequency and then triangulate them based upon the intensity of the signals. It looked easy to use. I switched it on again and quickly found the Wi-Fi receiver which was on the wall in the kitchen and seemed to be the reason that the unit had gone mad when I first used it.

I found the switch on the wall unit and powered off the wifi sender. Then there was a second source, coming from my iPhone. I soon switched that off and a third slightly weaker signal was still running. I triangulated it and discovered that it was coming from the flat screen television. Behind the television. I looked and there was a small black, flat unit wired into the TV. I assumed it was something to do with reception of pictures, except I could not see any direct connection from the device to the TV, except for its power connector.

I snapped a photo of the device on my iPhone and sent it picture to Bigsy - I'd expect to hear from him the next day.

However, it rang alarm bells with me. There appeared to be someone monitoring my room. I switched the WiFi back on and found myself tiptoe-ing around, quite unsettled.

M & A

Tuesday and I'm in the lab. Simon missed the bus this morning, but I didn't think I should call him to drag him out. I was with Schmiddi and Rolf brainstorming a new interface design and absent-mindedly mulling over what I could say to Juliette, when she walked into the lab. She was wearing her lab coat, and I realised she had been conducting some experiments.

"Hey Juliette," said Rolf, "You look happy today, almost as happy as our English man."

She looked across to us both but realised I wouldn't have said anything to Schmiddi and Westi.

"I've been with a couple of Selexor candidates today," she said, "I thought you'd all be interested."

"Should we wait for Amy?" I asked.

"I don't think so," said Juliette, "It could put her in a potentially disloyal position. I figured out how part of the systems works. Not Levi's original design, but the add-ons. The company using Selexor must train the system on what to look for and then tailor the test to a specific job. In effect, it is as if the employer's current workers filling the same job all sit through the same AI assessment. That's the entire spectrum, from high to low achievers"

Juliette continued, "Their responses are then matched with a 'benchmark of success' from those workers' past job performance, like how well they had met their sales quotas and how quickly they had resolved customer calls. The best candidates end up looking and sounding like the employees who had done well before the prospective hires had even applied."

She added, "Then, when a new candidate takes the Selexor test, the system generates a report card on their 'competencies and behaviours,' including their 'willingness to learn,' 'conscientiousness & responsibility' and 'personal stability,' the latter of which is defined by how well they can cope with 'irritable customers or co-workers.' Those computer-estimated personality traits are then used to group candidates into high, medium and low tiers based on their 'likelihood of success.'"

"So, do they immediately throw out some candidates?" I asked.

Juliette answered, "Not exactly. Employers can still pursue candidates ranked in the bottom tier, but they mostly focus on the ones the computer system liked best."

"But do we know how it works?" asked Rolf, "I mean,

how they have linked the Createl system to Selexor?"

Juliette shook her head, "Selexor offers only the most limited peek into its interview algorithms, both to protect its trade secrets and because the company doesn't always know how the system decides on who gets labelled a 'future top performer.'

Rolf looked dismayed, "So we can't find out how Createl links in, let alone how it influences Selexor outcomes?"

Juliette shook her head again, "Selexor has given only vague explanations when defining which words or behaviours offer the best results. For a call centre job, the company says, 'supportive' words might be encouraged, while 'aggressive' ones might sink one's score. They said its board of expert advisers regularly reviews its algorithmic approach, but the system isn't available for an independent audit."

"I wonder who their PR firm is?" asked Hermann.

"Look at this! Did you know that Createl was a Raven subsidiary - From this week?" said Rolf, looking quite pleased with himself. He put the new Press Release onto the big display screen, and we could all read it.

'Raven Corp acquires Createl

The massive multinational Raven Corp has acquired the remaining equity stake in the software vendor Createl, for an undisclosed sum. Raven has been an investor in the vendor for two years and reportedly held as much as a 70 percent stake before this deal. This will halt any rumours of an IPO for Createl. '

"Why on earth are we finding out about this through Press Releases?" asked Juliette, "Surely someone should be running a briefing?"

"I suppose they have kept it secret because it was commercially sensitive," said Hermann.

Rolf flipped further through the on-screen Press Release:

Createl specialises in AI analytics software, particularly focused on industry verticals and increasingly, shifting to the cloud with its AdVisor product. It has a solid, loyal customer base when sold integrated with the Selexor product

'Software is no longer an industry vertical; it is a disruptive layer that is transforming every facet of society,' said Jason Petronius, executive vice president and CEO of enterprises for Raven Corp.

'As a global organisation spanning multiple industries across 60 countries, Raven has the resources, knowledge and relationships to help Createl continue to expand its transformative capabilities.

'Createl has invested approximately $200 million in product design and development to deliver industry-specific AdVisor suites that solve the most challenging staffing issues for customers around the globe. Createl's AdVisor suites are cloud-native, industry-specific and built to help companies modernize and drive immediate, tangible impact.

'Createl, when used with Selexor operates in key industries including: manufacturing, distribution, healthcare, public sector, retail and hospitality.

'Sandgate Road Capital founded and began building Createl five years ago,' said Lazlo Francis, co-founder of Sandgate Road Capital. 'We are selling our remaining stake to Raven Corp, our partner in Createl for the past three years, because of the significant strategic value between Createl and Raven and its subsidiaries, particularly Selexor.'

Sandgate Road Capital Managing Director Rajendra Dhavale added: 'Over the past years working with Raven Corp and the management team, it is very apparent that there is a great partnership with Raven that will take the company to its next level of success. We are very proud to have worked with the management team in transforming Createl repeatedly over the years as the market has evolved.'

'Ravens's decision to acquire Createl is a strong endorsement of our product strategy and focus on creating innovative solutions for our customers,' said Mary Ranzino , CEO of Createl in a statement.

'As a subsidiary of a multi hundred billion dollar revenue company that re-invests 90 percent of earnings back into its businesses, we will be in the unique position to drive digital transformation in the markets we serve.'

Mary Ranzino added, 'We are rapidly expanding our industry-specific AdVisor and offering client recruitment experiences and outcomes that are well beyond what is standard in enterprise solutions. This is achieved through our close operation with Brant Industries' Selexor, another of Raven's subsidiaries.'

At the hotel and Casino giant Logan International (LI), thousands of applicants for reservation-booking, revenue management and call center positions have been provisioned through Createl and Brant's Selexor AI system, and executives credit the automated interviews with shrinking their average hiring time from six weeks to five days.

Diana Charles, the Logan International's vice president of global recruitment, said the system has radically redrawn LI's hiring rituals, allowing the company to churn through applicants at lightning speed. Hiring managers inundated with applicants can now just look at who the system ranked highly and filter out the rest: "It's rare for a recruiter to need to go out of that range," she said.

At a well-known consumer goods conglomerate, Createl is credited with helping save 100,000 hours of interviewing time and roughly $1 million in recruiting costs a year. Selina Powers, the company's chief human resource officer, said the system had also helped steer managers away from hiring only

"Mini-Mes" who look and act just like them, boosting the company's "diversity hires," as she called them, by about 16 percent.

'The more digital we become, the more human we become,' she added.

Dane E. Persillor, the global head of human-capital management at Createl and Selexor client Prymordial, wrote in the Planet Business Review this spring that the banking giant's roughly 50,000 video-interview recordings were 'a treasure trove of data that will help us conduct insightful analyses.'

Prymordial said it uses Selexor's video-interview system but not its computer-generated assessments. But Persillor said data from those videos could help the company figure out how candidates' skills and backgrounds might correspond to how well they would work or how long they would stay at the firm. The company, he added, is also "experimenting with résumé-reading algorithms" that would help decide new hires' departments and tasks.

"Can I imagine a future in which companies rely exclusively on machines and algorithms to rate résumés and interviews? Maybe, for some," he wrote. (The "human element" of recruiting, he pledged, would survive at Prymordial.)

Createl's expansion has also helped it win business from smaller groups such as Bottle:IT, a Chicago nonprofit organisation that trains unemployed local job seekers for careers in the tech industry. Crispin Waller, the group's chief, said Selexor with Createl had proved to be an irreplaceable guide in assessing which candidates would be worth the effort.

The nonprofit organisation once allowed almost anyone into its intensive eight-week training program, but many burned out early. Now, every candidate goes through the AI assessment first, which ranks them on problem-solving and negotiation skills and helps the group determine who might have the most motivation, curiosity and grit.

"Knowing when that person is at a starting place, when it comes to this person's life," Waller said, "can help us make more accurate assessments of the people we're saying yes or no to."

"Look and some alternative commentary, " said Rolf, gleefully:

Diana Tensman Batenock, a neuroscientist who studies emotion, said she is 'strongly sceptical' that the system can really comprehend what it's looking at. She recently led a team of four senior scientists, including an expert in 'computer vision' systems, in assessing over 1,000 published research papers studying whether the human face shows universal expressions of emotion and how well algorithms can understand them.

The systems, they found, have become quite perceptive at detecting facial movements — spotting the difference, say, between a smile and a frown. But they're still worryingly imprecise in understanding what those movements actually mean and woefully unprepared for the vast cultural and social distinctions in how people show emotion or personality.

'Look at scowling,' Batenock said: 'A computer might see a person's frown and furrowed brow and assume they're easily angered — a red flag for someone seeking a sales associate job. But people scowl all the time,' she said, 'when they're not angry: when they're concentrating really hard, when they're confused, when they have gas.'

"Hilarious," said Rolf, "Now they can de-select someone who has eaten too much chilli the night before their interview!"

My mind was racing now. It it looked as if Createl had been sold to Raven and Selexor was already part of Brant, a Raven subsidiary. Now Raven owned the entirety of the AI selection software suite.

Juliette began, "I don't think we'd better tell Amy any more of the thoughts that some software might be 'snake-oil,' she said.

"Amy knows," said Hermann, "We've talked about it often enough in the office and it was also well reported after (sorry, Juliette) Levi's death.

Rolf chipped in, "It seem to be more the people higher up in the company that are trying to spray over these inconvenient facts. And did you notice? I hadn't realised it but the Managing Director of Createl - Mary Ranzino - shares the same name as one of the Senior Vice Presidents of Brant, Bob Ranzino?

He tapped away and found an article: Bob and Mary

Ranzino host a gala Veteran's Party from their Byram Shore upscale home in Old Greenwich, Connecticut.

Rolf put the pictures on the screen. The great and the good of Connecticut were displayed at a lavish fund-raiser, in a sumptuous waterside location.

"It's got its own pool and jetty alongside the house!" said Hermann. "Like something out of Great Gatsby!"

"Yes, and it is in Old Greenwich too, which I think is the mansion-intensive area of Greenwich!" said Rolf. The house pictures in the background showed a clapper-boarded home complete with turrets to round off the corners.

"I thought an Englishman's home was his castle?" said Hermann, laughing and looking towards me.

"Well, it is New England, I suppose" I quipped.

Quayside at Bel-Air

Tuesday evening and I rode back on the bus to Bel-Air. There was still no sign of Simon, so I assumed he'd had to skip the day. I'd call in on him during the evening.

Once I was at Bel-Air, I decided it would be a good time to call Amanda. It was difficult to do so from the Apartment, because of the chance of being overheard. It was also a challenge from Bel-Air because it is one of the major transportation hubs in Geneva. There are trams and buses tipping people all over the pedestrian precinct. I decided I'd take a route along the river instead of the way most people were walking and found myself on a piece of quayside that didn't seem to go anywhere.

I resigned myself to making one of those walk-about phone calls on a headset, when I rang Amanda.

"Hello Matt,"

"Hi Amanda, I'm checking in. We've a few developments here."

"That's good. Are you outside of your apartment? It sounds like you are in the street?"

"Correct - I'm worried that my Apartment has a listening device installed. Bigsy sent me a scanner, and it has picked up something. I don't know what is best, but I thought I should contact you away from the Apartment."

"Good. I'll get someone to look into it for you. I'm sure we can get whatever it is removed."

I continued, "Other news. Raven have just acquired Selexor, which is the AI company that uses the Brant product Createl, which we are all working on. It means the whole system is now owned by Raven. Oh, and there's a strong feeling from the developers that Levi Spillmann's invention worked well for bug spotting in fields but would not be applicable to human use."

"Yes, I've seen in various reports that the system was extremely good at finding crop infestations but that the AI used would not be applicable elsewhere."

I added, from the morning's briefing, "Well, Selexor has been very successful at selling itself into major corporates for employee filtering - and they use the Createl AI scanning as part of the process."

"Some critics say it is smoke and mirrors? That really Selexor doesn't work?"

I offered my opinion to Amanda, "Yes, although I have a sneaking suspicion the filter questions that Selexor uses are to weed out most of the deemed non-desirable candidates. For example, the question logic can filter out, say, those with less than a 2.1 degree. Those with a BMI of over 25 or under 18, Those with a passport from a country ending in 'stan'. Those with a surname ending in 'witz'. You get my drift."

Amanda spoke, "If so, it is truly shocking and wholesale discriminatory."

"Yes, and they don't reveal the algorithm, nor the reasons for a candidate rejection. They say it is proprietary intelligence. And then they wrap it up with references to Createl's Artificial Intelligence."

"I guess they told you that there is a small cloud over Simon Gray?" I asked.

"Yes, as a matter of fact, Jake told me. Apparently, Christina and then Clare spotted something. He appeared to have a Russian prison tattoo and Christina overheard him speaking Russian into a cell phone. I've asked Christina to fly over, to check him out. She'll be staying in Four Seasons Hotel des Bergues, from tonight. It's not that far from your Apartment."

"I know it. It's just along from the Mandarin Oriental. Very central actually."

I decided that it would be a good idea to catch up with Christina that evening, if she was in town, ideally before she met Simon. Maybe I could catch up with Simon tomorrow instead of bothering him this evening.

"The whole situation around Levi Spillmann is also quite

complex," I said to Amanda, "He invented the original algorithms whilst he lived in Israel. His girlfriend of the time was arrested, incarcerated, and then allegedly hanged herself in prison. I'm not convinced. Spillman then came to join Brant, bringing the Createl algorithms with him. He became the boyfriend of Brant employee Juliette Häberli, before he died in a boating accident. A boat on which Simon Gray also occasionally crewed. Oh, and Simon was a crew member on fancy yachts out on the Barents Sea, in his past."

"It's making my suggestion that we get Christina to check him out seem all the more necessary," said Amanda.

"I think we should check whether Juliette is on the level too," I said, "I mean she seems genuine, but this whole thing is so mixed up."

"You have a number for Christina? I'll call her and arrange a meeting," I said.

"It's on its way," answered Amanda.

Christina in Geneva

Now I was in the Four Seasons, waiting in the plush blue chairs of the bar for Christina. She had agreed to see me that evening and arrived a few moments later looking like a rock star. An off-white tee shirt top, with the sleeves pulled up by a couple of epaulettes, a gold flash across her chest, long hair down and a pair of ray-bans being worn like a headband. And another pair of those impossibly tight trousers, this time in a black fabric.

"Hej!" she said and didn't do any of the complicated bise kissy-kissy moves, but instead just planted one on my left cheek. She hugged me as well, and I could somehow feel the energy from her body. I'd got a table in the corner, near to the bar and the window.

"Any preferences on where to sit?" I asked.

"No, this is good, and we have control of the room from here," she said, "It's been a busy evening actually - I was with my friend Elena earlier, she's from Milan but has been visiting a client here in Geneva. She came by train, the flight goes via Roma, or Amsterdam! Madness, isn't it!"

"How did you know she'd be in Geneva?" I asked, "Oh, we were chatting last week. I remembered that she worked for Brant, and I thought I'd pick her brain. As a matter of fact, they asked me to investigate something to do with Raven and Brant a while ago, and it is how I came to meet Amanda and Chuck."

I was thinking to myself that I must lead a boring life compared with Christina, who seemed to have useful friends everywhere. Still, this time I was the one she was visiting in the foreign country.

"Have we found out anything more?" asked Christina, "I've made a few enquiries since we were all in London. Raven now owns both Selexor and Createl. The person who hired you, Bob Ranzino, is married to Mary Ranzino who is the Managing Director of Createl. Most of the trade press are sceptical of Selexor's claims that it can filter employees using Artificial Intelligence. This isn't the first time that a Brant Company has made false claims either. They did it with Elixanor, when they used a 19-year-old female CEO to make false claims about the healthcare benefits of the product set."

I was a little phased by this quick brain dump by Christina. She seemed to have found out just about everything that I had discovered, and a few things more, but seemed to have done it from London. I wondered exactly how she was connected to the Intelligence Services.

She continued, "I want to meet Simon Gray. Back in

London, I thought what a name. Everyone knows that John Smith and David Jones are common British names, but usually the name Simon Gray gets missed, like Ian Black.

"But you check it out. Type Simon Gray into a search engine and there will be dozens, with all manner of professions. It is a great name to use to create a search for a needle in a haystack."

She pulled out her phone and demonstrated:

"See," she leaned over, and I caught her Chanel fragrance, "See: Counsellor, pool player, weight watcher, solicitor, footballer, builder, artist, career counsellor, digital transformation expert, architect. Now watch as I hit scroll:"

She did, and the list continued: lifeboat crew, photographer, marketeer, author, rugby player, businessman, sales recruiter, eBusiness consultant, sales engineer, business manager - perfumes, maritime engineer.

"And that's just if I use LinkMe" she said.

"I see," I said, "It is a great name to hide behind."

"Precisely. So, what have we got? An anonymous identity, referred to Amanda for this mission. Polished English - Claims to be from Hampshire yet no real sense of London's geography. He's got prison star tattoos that say, 'don't mess with me,' and he speaks Russian. Does this remind you of anything?"

"Er - a plant? A Russian plant inside MI6?"

"Correct, but more than that. A GRU cutout inside Amanda's operation."

"Why do you say GRU?" I asked.

"The Main Directorate of the General Staff of the Armed Forces of the Russian Federation, officially abbreviated G.U."

"I though the GRU was shut down in the 1990s?" I asked.

"It was, but the GU is still usually referred to as the GRU, like SI6 is still called MI6" answered Christina, "GRU controls the military intelligence service and maintains its own special forces units."

"Isn't this somewhat overkill for the situation?" I asked.

"Not really, unlike Russia's other security and intelligence agencies—such as the Foreign Intelligence Service (SVR), the Federal Security Service (FSB), and the Federal Protective Service (FSO), whose heads report directly to the president of Russia—the director of the GRU is subordinate to the Russian military command, reporting to the Minister of Defence and the Chief of the General Staff.

Christina continued, "Some say this gives it a clandestine power that the other agencies don't have. Consider Russia as a kleptocracy. The oligarchs seized power and money and nowadays run the state for their personal gain. It is convenient to have a powerful spy network on-side. The directorate is reputedly Russia's largest foreign-intelligence agency, and distinguished among its counterparts for its willingness to execute riskier complicated, high-stakes operations.

She added, "It does that by sometimes hiring dubious agents, such as prisoners who don't have so much to lose. They can be converted to hitmen or to Spetsnaz forces who won't ask too many awkward questions about their mission."

"I'm not sure I like where this is leading," I said, "The evidence is starting to imply that Simon is a Russian GRU member embedded inside the Brant organisation. More that he could even be a hitman, recruited from prison?"

"I have heard of stranger things," said Christina, "The twist here is that he speaks such very good English, but then I suppose it depends how much attention one pays to accents and dialects."

"Quite a lot," I replied, "For example I can tell that you are a well-heeled Londoner, who has travelled abroad and has additional languages."

"Not even close," smiled Christina, "Actually I'm from Reykjavik, Iceland."

I had to hide my head in shame after that. I would never have guessed.

"So perhaps it would be more natural for me to meet Simon with you?" suggested Christina, "He's seen me before, when he came to London, but you could say that I'm here in Geneva checking on the Raven takeover. We don't want to do anything to make him suspicious."

"Why would you call me though?"

"I got your number from Amanda but didn't have his and I didn't want to ask Clare."

"Okay, I think that should work; we should do this tomorrow. I know a great lakeside spot for us all to meet."

"La Terrasse?" inquired Christina.

"Oh, you are good. Very good," I replied.

"Okay - Thanks for all of this," she said, "I'll be going now, busy day tomorrow!"

She stood, and I did, she kissed and hugged me softly, "Be careful," she said.

Walking back from the Four Seasons, I decided not to call Simon until I was back at Bel-Air. Otherwise, if he said he'd be able to see me straight away, I would take an unrealistically long time to walk back. I didn't want his fancy listening device to have something else that could work out where I was calling from.

It was all for nothing though. I got through to his mobile phone, but then had to leave a voice mail. I said that Christina would be in town tomorrow and it would be great for us all to go for a drink after work. I suggested

La Terrasse at 18:00.

Then I texted Christina and let her know the plan.

Simon misses the bus

Wednesday morning and Simon wasn't on the bus again. It was bugging me a bit now, particularly as I'd arranged for him to attend a session today in our Lab. It was being presented by Juliette and Hermann and was about BCI (Brain Computer Interface) technologies and ways to improve the linkages. I thought that Simon's work on the chemistry of neural pathways would be an interesting addition to the session and so I'd invited him along.

On the way in, on the bus, I decided to call Simon's newest neighbours, Bradley and Jennifer.

Jennifer picked up: "Hello, this is Jennifer Hansen."

"Hello Jennifer, this is Matt - the new addition to the Apartments. We met last week at the 'plum pie party'."

"Oh hello, Matt, yes I remember you - very well, actually. You just surprised me calling at this early hour. Is everything okay?"

"Oh yes, thank you Jennifer, everything is fine. I was

wondering about Simon Gray, your near neighbour. It's just that I haven't seen him for a column of day and he normally catches the same bus to work as me. We both work at Brant and it is a company bus."

"Oh yes, you are both scientists! I remember. No, I've not seen him, not in the hallways either. Shall I ask Bérénice Charbonnier?" she asked.

"Bérénice?" I queried, "Oh, you must remember her? She was teaching you about faire la bise! But she is also Simon's girlfriend."

I had no idea, and he had never mentioned her.

"Oh, yes please, would you? You've got my number here. It would put my mind at ease to know that everything is good with Simon."

"Oh certainly," she replied, it's good when neighbours look out for one another."

I said my thanks and by that time we were almost at the campus park. We drove in and I disembarked from the bus. My phone tinged and it was a message from Jennifer."

"Bérénice also concerned. Has no Apartment key so we asked Aude Darmshausen to bring around a key. Aude to drive around at 18:00."

I wondered how much of a storm in a teacup I'd now created. At least I'd discovered that Simon's girlfriend was Bérénice, but I'd also accidentally alarmed her and dragged the Aude the Courtière Locations Résidentielles around as well. Now I'd have to rearrange things with Christina, because we"d agreed to meet at 18:00 at La Terrasse."

I sent her a text too, explaining the change of plan, and she sent back a 'thumbs up' okay signal. I wondered if she would come to the Apartment for six pm. Right now, I had to concentrate on the Juliette and Hermann show, in the main auditorium.

Pipe

"You are sitting and smoking;

you believe that you are sitting in your pipe, and that *your pipe* is smoking *you*;

you are exhaling *yourself* in bluish clouds.

You feel just fine in this position, and only one thing gives you worry or concern:

how will you ever be able to get out of your pipe?"

— **Charles Baudelaire, Artificial Paradises**

Mind Meld

Amy was also in the Auditorium and sat next to me for the presentation.

"Some of this is getting out of control, " she whispered, "I'm concerned that Kjeld is asking us to do things which we know are impossible, using sometimes flawed technology. Let's see what Juliette and Hermann have to say."

Juliette walked on to the stage of the auditorium. She was at ease running a presentation in such surroundings. She was in a presentation-ready outfit, unlike her daily Lab garb. Pastel pink jacket, white hoop-necked undershirt, and a pink tight-fitting skirt. Schmiddi was in one of his regular suits and a very slightly crumpled looking shirt with a grey tie, which was probably expensive. They both wore small badges in their lapels. Juliette's looked more like a brooch and Schmiddi's identical one looked like he belonged to a political party.

Juliette started with a picture from Star-Trek. It was of Spock, the half man, half Vulcan, gripping a man's face, where the man was pressed against a wall.

Juliette began, "The Vulcan mind-meld originates in the Star-Trek episode 'Dagger of the Mind,' in which Spock probes the tortured mind of a prison inmate to get to the bottom of a confounding mystery about people going crazy from a machine that is supposed to be healing them.

"We must remember that Spock is only half-Vulcan because his mother, Amanda Grayson is a human.

"We know that Star Trek was a TV show, but we must not let ourselves get carried away with the devices and concepts it used. Whatever magic brain energy is necessary to do the mind meld, should it work with a machine?

Hermann stepped forward, "As we say, Wetware isn't Hardware."

Then Hermann said, "Now let's look at the Wetware to Hardware interface. As ancient Greeks fantasised about soaring flight, today's imaginations dream of melding minds and machines as a remedy to the problem of human mortality. Can the mind connect directly with artificial intelligence, robots and other minds through brain-computer interface (BCI) technologies to transcend our human limitations?"

I was impressed with their two-handed slickness. Hermann continued, "Over the last 50 years, researchers at university labs and companies around the world have made progress toward achieving such a mind-meld vision. Brant's Selexor used Createl and other entrepreneurs such as Elon Musk with Neuralink and Bryan Johnson with Kernel have visionary companies that seek to enhance human capabilities through brain-computer interfacing.

"Now it should be said that much of the recent work on Brain Computer Interfaces aims to improve the quality of life of people who are paralysed or have severe motor disabilities. University of Pittsburgh researchers use signals recorded inside the brain to control a robotic arm. Stanford researchers can extract the movement intentions of paralysed patients from their brain signals, allowing them to use a tablet wirelessly. Some limited virtual sensations can be sent back to the brain, by delivering electrical current inside the brain or to the brain surface.

Juliette stepped forward and asked, "What about our main senses of sight and sound? Very early versions of bionic eyes for people with severe vision impairment have been deployed commercially and improved versions are undergoing human trials right now. Cochlear implants have become one of the most successful and most prevalent bionic implants–over 300,000 users around the world use the implants to hear.

As she spoke, the slides were changing to illustrate these early trials. Black and white photos emphasised the early primitive nature of these trials.

Hermann added, "A bidirectional brain-computer interface (BBCI) can both record signals from the brain and send information back to the brain through stimulation. The most sophisticated BCIs like those developed here in Brant are "bi-directional" BCIs (BBCIs), which can both record from and stimulate the nervous system.

Herman chose to illustrate it with a simple diagram of a head and a computer chip with arrows between them, pointing in both directions.

He continued, "We're exploring BBCIs as a radical new rehabilitation tool for stroke and spinal cord injury. We've shown that a BBCI can be used to strengthen connections between two brain regions or between the brain and the spinal cord and reroute information around an area of injury to reanimate a paralysed limb.

Juliette stated, "With all these successes to date, you might think a brain-computer interface is poised to be the next must-have consumer gadget."

She flipped up a picture of a wiring grid, with three sets of 16 probes on it and said, "This is a picture of an electrocorticography grid, used for detecting electrical changes on the surface of the brain. And it is being tested for electrical characteristics. Look around the other side of it and see the sharp pins which can be dug into the brain for these sensors to function.

Again, a simple diagram, which looked like a chip with a set of around 50 needles poking out of it.

And then Hermann spoke again, "A careful look at some current BCI demonstrations reveals we still have a way to go: When BCIs produce movements, they are much slower, less precise and less complex than what able-bodied people do easily every day with their limbs."

"For example, bionic eyes offer very low-resolution vision; cochlear implants can electronically carry limited speech information, but distort the experience of music. And to make all these technologies work, electrodes have to be surgically implanted – a prospect most people today wouldn't consider. Literally brain surgery. We have some of the technology, but it is still rudimentary."

Juliette added, "Not all BCIs, however, are invasive. Non-invasive BCIs that don't require surgery do exist; they are typically based on electrical (EEG) recordings from the scalp and have been used to show control of cursors, wheelchairs, robotic arms, drones, humanoid robots and even brain-to-brain communication. But most people can see that any form of inductive link will not be as precise nor as powerful as something directly connected to the brain."

She showed a picture of an electric toothbrush, "See, one of the first inductive loops, used to charge an electric toothbrush. Years later, it is still much slower than a direct connection into the toothbrush."

She continued, "The first demonstration of a non-invasive brain-controlled humanoid robot avatar named Morpheus was in the Neural Systems Laboratory at the University of Washington. The non-invasive BCI infers what object the robot should pick and where to bring it based on the brain's reflexive response when an image of the desired object or location is flashed."

Hermann added, "All these demos have been in the

laboratory – where the rooms are quiet, the test subjects aren't distracted, the technical setup is long and methodical, and experiments last only long enough to show that a concept is possible. It's proved very difficult to make these systems fast and robust enough to be of practical use in the real world. This work is difficult."

He continued, "Even with implanted electrodes (ie brain surgery) another problem with trying to read minds arises from how our brains are structured. We know that each neuron and their thousands of connected neighbours form an unimaginably large and ever-changing network. What might this mean for neuro-engineers?"

"Imagine you're trying to understand a conversation between a big group of friends about a complicated subject, but you're allowed to listen to only a single person. You might be able to figure out the very rough topic of what the conversation is about, but not all the details and nuances of the entire discussion. That's the challenge. Even our best implants only allow us to listen to a few small patches of the brain at a time. We can do some impressive things, but we're nowhere near understanding the full conversation."

Juliette continued, "There is also what we think of as a language barrier. Neurons communicate with each other through a complex interaction of electrical signals and chemical reactions. This native electro-chemical language can be interpreted with electrical circuits, but it's not easy. Similarly, when we speak back to the brain using electrical stimulation, it is with a heavy electrical 'accent.' In other words, we can't yet speak 'Brain

Language' very well. This makes it difficult for neurons to understand what the stimulation is trying to convey in the midst of all the other ongoing neural activity."

Hermann added, "Finally, there is the problem of damage. Brain tissue is soft and flexible, while most of our electrically conductive materials – the wires that connect to brain tissue – tend to be very rigid. This means that implanted electronics often cause scarring and immune reactions that mean the implants lose effectiveness over time. Flexible biocompatible fibers and arrays may eventually help in this regard. We were hoping to get some thoughts from our contacts working on the chemistry of neural pathways, and maybe this will follow."

I realised that this was a sideswipe at the non-appearance of Simon, nor of anyone from his lab.

Juliette displayed the latest headgear experiments. She showed a projection of the headgear and then produced a sample from a box. It was like a cycle helmet, with an internal array of miniature shower head nozzles and a large wire snaking away from it.

"We used to call it the ServoCask, from casque de cerveau but then Brant asked us to change it to Cyclone."

Juliette spoke, "This Cyclone headgear records brain signals without the need for surgery and can either measure the electromagnetic fields generated by groups of neurons or detect small changes in blood oxygenation, which correlate well to nearby neural activity."

She added, "We are using magnetometers to measure tiny changes in magnetic fields and light pulses through the skull and into the bloodstream in order to measure how much oxygen the blood is carrying at any given time."

Juliette continued, "It's a similar concept to the way that smart watches measure blood oxygenation but has a vastly extended coverage. The headgear takes advantage of the relative transparency of the skull and brain tissue to near-infrared light by beaming photons through the skull and measuring their scattering and absorption, allowing inference about blood flow and oxygenation. That's something called haemodynamics."

Hermann added, "This headgear offers the resolution and sensitivity of state-of-the-art haemodynamic systems across the top layers of cortical tissue.

Herman produced an animated diagram, "Traditional 'continuous wave' near-infrared spectroscopy devices apply light to the head continuously, which then scatters throughout and is detected at various locations upon exiting the head. It has a decent level of accuracy but the processing time (we'd call it latency) of the system means it is like having very slow reactions. In fact they are more like the reactions from someone on the other end of a satellite phone call."

I was thinking about some of Brant's early uses for the tech. They wanted to deploy it in some military scenarios. It would be like using slightly dim low-fidelity mogadon-zombies in a war-zone. Hopeless.

Hermann continued to explain, "Changes in the detected light intensity allow inference of optical-property changes inside the head, like those resulting from neural activity. An analogy would be using a sonar device to detect the movement like shoals of fish in a lake."

I thought this all sounded too slow and only good for tracking big objects.

Hermann put up a new diagram, "Time-domain systems capture a much richer signal by applying the light in short pulses and precisely capturing the arrival time distribution of scattered photons for each pulse. On average, photons that arrive later travel deeper through the tissue, which reveals additional depth-dependent information about the optical properties of the tissue - allowing for more detailed inference of brain activity."

Inference - there was the word again. It was still mainly guesswork.

Hermann triumphantly concluded, "In other words, this Cyclone headgear is about the most advanced non-invasive haemodynamic and photon inference machine-to-brain communication device anywhere."

Juliette added, "With all these challenges, we're very optimistic about our bionic future. BCIs don't have to be perfect. The brain is amazingly adaptive and capable of learning to use BCIs in a manner similar to how we learn new skills like driving a car or using a touchscreen interface. Similarly, the brain can learn to interpret new types of sensory information even when it's delivered non-invasively using, for example, magnetic and light pulses.

The end of the presentation had a Question-and-Answer session, but without anyone from the CNP team (chemistry of neural pathways team) it was not particularly interesting. I was intrigued about how much the brain would have to adapt to using the headgear. I recalled the way that eyes adapted to wearing spectacles, where the brain would average out the improved vision, but it would often take a few days. Then I wondered if everyone wearing the Cyclone would need a haircut first.

Amy looked towards me and I said, "You know, that is an amazing leap forward in the technology, but if the AI systems are not good enough, then there is still a way to go."

Amy said, "Yes, my thoughts too, a chain and its weakest link. We have several good building blocks here but some vital pieces are missing."

I was thinking that it was still a case of join the dots, but someone had erased some of the dots too.

Inference. Latency. Shoals of fish. Haircuts. And that was before the system was hooked up to the shonky AI system. As the French-speaking Swiss might say, "Pah!"

Bérénice has a bit of a reputation

Wednesday evening and I needed to get back to the Apartment to meet Christina and Aude. Juliette came to my rescue at the exit from the lab.

"You are looking anxious," she said.

"Amy asked to see me this evening and now I'm running late. I'm supposed to be meeting my Apartment Concierge and a friend at 18:00, and now I'm running late."

"Don't worry," Juliette answered, "I'll drive you; I fancy going into Geneva tonight in any case."

"Brilliant presentation, today, by the way!" Juliette was still in her 'Presentation clothes' from earlier and looked like she was one of the chief executives of the company.

She noticed me looking up and down her clothes. I felt I'd need to say something.

"Oh yes, stunning outfit!"

"I wanted to look good as I was presenting to so many people!" she answered.

One of the people we are to see this evening is Simon Gray, he is still in the apartments. That won't be awkward for you? Will it?" I asked.

"No," she said," he looks as if he feels awkward around me, but I assume it's because he doesn't know what to say about Levi. I think he has a girlfriend in Geneva. One of the women who lives in the same Apartment block, I believe."

"Oh, yes, I'm meeting her too. Bérénice is her name.

"Not Bérénice Charbonnier?" asked Juliette, "She was living in the Apartments when I was there. As a matter of fact, she had a bit of a reputation - las respectueuse."

"Nope," I said, "It doesn't ring a bell."

"Okay the literal translation, dame de la nuit,"

"Oh," I said, "I can work that one out."

"Your French is getting better every day, " said Juliette.

We arrived at the Apartment and Juliette parked the car. We climbed out just as Christina appeared from across the street wearing a short leather biker-style jacket, a grey and white raggedy looking scarf and yet another pair of tight-fitting, this time slightly shiny trousers. I decided she was carrying on with the rock-chick look.

"Nice car," she said smiling. I introduced Juliette and Christina started a lively conversation in French. I think it was something about Juliette's clothes, because Juliette showed the jacket and then turned around in the skirt. Then Christina did the same in her outfit. They both laughed and I could see they were getting on well.

"Okay, Let's get to Simon's floor," said Christina.

We used the elevator and arrived to see the door to Simon's room opened. I tapped lightly on the door and poked my head inside.

"Simon?" I called.

"He's not here," answered Bérénice, looking towards us all as we entered the room. Aude was also standing in the living room.

"Oh - Hello Juliette! It's been a while since we saw one another." They switched to French but didn't greet one another in the French way, and I assumed their chatter was about Simon.

Aude spoke next, "It is very unusual, Simon has been a long-standing resident here, but now it looks as if he has gone. Bérénice says that all of Simon's things have gone, although she says that some of hers are still here. Mainly a few clothes in the wardrobe. But for Simon there is nothing left, not even a toothbrush."

Christina walked across, "Bérénice has not been here for a couple of days, but also couldn't contact Simon. She said it wasn't that unusual for Simon to go quiet for a few days occasionally, and it was mainly when Matt called Jennifer that she started to get worried. Matt said Simon was due to be at a special presentation today in Brant."

Aude asked, "Look, I don't know what any of this is about, but would you like me to call the police? Do you suspect something irregular?"

"No, no," answered Bérénice, "I'm sure this will have a simple explanation. I'll try to phone him again in a minute. If there's anything unusual, we will call the police. I am just so used to this happening with Simon."

"But isn't it unusual that all of his clothes have gone?" asked Aude, "It looks as if he has suddenly moved out."

Christina was looking around the room. She noticed a sunken floor socket in the living room and opened the cover.

"There we are. A couple of trailing wires! He's forgotten to take the chargers" She flipped open the cover and we could all see a small mains adapter with Xaomi written on it and a tiny neon green USB device like a memory stick protruding.

"We need Bigsy for this," said Christina, as she removed the tiny unit from its power supply.

"Who are you all?" asked Bérénice.

"Oh, we're just friends of Simon," answered Christina, "I knew him from London, but not so well."

I'd been looking at the small green unit sticking out of the mains adapter. I recognised it as a Wi-Fi repeater stick. With some luck, we could interrogate the settings on it to get some more information.

"Look," said Aude, now speaking in her clear home counties English, "I'm not sure what this is about but I'll trust all of you not to do anything to Mr Gray's apartment and to call the police and notify me if there is anything untoward. But I'm going to leave now. Mr Nicholson? Can I have your word that you will alert me if you discover anything unusual?"

I agreed and even said I'd let Aude know the outcome whatever happened, and I thanked her for letting us into the Apartment.

"Check the kitchen drawer," she whispered, "I give everyone the same advice about keys."

Sure enough, there was a spare key to Simon's apartment in the kitchen drawer.

Aude left, I quietly retrieved the key, and we all left the apartment. I noticed that Bérénice didn't look too concerned as we walked back to the elevator.

"I've got your number now, " I said, "I'll call you if I hear anything,"

Bérénice nodded, "I'm only along the hallway," she said.

Christina, Juliette, and I walked to the elevator and then

down to the outside of the building.

Juliette was suitably discreet around Christina and we all walked back to Juliette's car.

"Wow - that was interesting, "said Christina, "What do you say we all go for a drink? My hotel is nearby, we could have a couple of drinks and try to work out what has just happened?"

"Oh, where are you staying? " Asked Juliette.

"The Four Seasons," replied Christina, "They have a good range of cocktails there."

"I can leave my car here, or even better I can park in the underground car park before we go, then we can grab a taxi back to your hotel," suggested Juliette.

"Good plan," said Christina, "Matt, can you get us an Uber or something?"

I called the local cab company, and they said they would be along in around eight minutes.

Juliette moved her car, and I stood with Christina.

"What will we tell Juliette?" I asked.

"Just nothing about MI5, but anything else is fine, even your links to The Triangle," answered Christina, "And we should try to find out whether she knows anything else. You think she is on the level?"

"Oh yes, I said, I'm sure of that!"

"Like you were sure of my accent?" asked Christina.

Terrasse, talking about pop music

The taxi arrived and we all jumped in. Christina asked, in French for the destination and we moved off.

"Great idea," said Juliette.

"I love that place, " said Christina.

"What?" I asked.

"We're going to La Terrasse," explained Christina, "I'm only in Geneva until tomorrow and I'd feel cheated if we could not get out for a while."

It was another pleasant evening, and the sun was still high enough to bathe us in good light as we sat at one of the tables. The same server came along and greeted Juliette. Then he looked at Christina.

"Non!" he said, "Christina Nott! I saw you in Montreux! Incroyable! I have your album. Please, can we have a picture together!"

I could work out what was happening from the words, but none of it made sense. The waiter seemed to think that Christina was a pop singer.

He called over another waiter and Christina posed for a couple of selfies, one with arms around the waiter and the second with Christina planting a kiss on his cheek.

When we were eventually asked what we would like to drink, I could see the poor waiter trembling with the adrenaline rush.

"I should explain," said Christina, "My other job is as a pop singer. I've toured this area as well as, well, the globe. My last tour was mainly in America, although we played Tokyo as well."

"Oh my god," said Juliette, "You are that singer with a string of hits about break-ups and mysterious assignations in Paris and that one about feeling your body all over mine. I get goosebumps when I listen to some of those tunes. Matt - you know some interesting people!"

Christina nodded. I was still none the wiser but reckoned that it accounted for the rock-chick chic of her dressing.

"But what about you?" said Christina, "You seem to be an interesting lady too? Clever according to Matt and very knowledgeable about cybernetics and psychology?"

"Yes, I'm surprised I spotted nothing unusual about Simon, really," said Juliette.

"But you did," I said, "You thought it unusual that he was around Levi as much, at the yacht, and it surprised you about his girlfriend, Bérénice,"

"Well, both of those are true, I suppose. To be honest, I thought he had a bit of a man-crush for Levi. It's why I was surprised about his assignation with the neighbourhood girl."

"Ew," said Christina. "That's a bit harsh about Bérénice, isn't it? And a bit of a strange thing to say about your boyfriend - oh my condolences as well."

"Thank you," said Juliette, "But Bérénice's story is well-known. I think she has been through every man that has stayed at the Apartments, even (en confiance), Bradley - that was a week when Jennifer was back in America for a special convention. I'm surprised she hasn't made a move on you yet, Matt.

"Well, she gave me a lesson on kissing the day I moved in," I said.

They both looked at me at that moment. I felt I'd overstepped an invisible, but to them blindingly obvious, line.

"Juliette, how did you get to be involved with Brant?" asked Christina, "I've had some run-ins with Raven in the past and they seem to be an unsavoury bunch of characters."

"I started out with Createl when it was a much smaller outfit. We all knew one another, and it was quite like a big family. Around the time when Levi joined, I could see that things were getting serious. Brant acquired all his intellectual property and by arranging a link up with another company we could move faster. I think we've messed that up though."

Juliette continued, "We've moved so fast that we've left the integrity of the product behind. Now we are promoting false promises to everyone.

"Take today, I had to run the session to show the newest capabilities of the HCI - the Human Computer Interface. I showed off the new Cyclone headgear which can be used to process brain signals and it tells a convincing enough story. Even with all of Hermann's and my caveats, by the time it has been spun by our Press Office it all sounds believable. Hermann and I were using an autocue system today. Our speeches - our very moves were scripted. We are just practised at making it look realistic."

"But have you sold your soul to the company?" asked Christina.

"It feels as if I have sometimes," said Juliette, "But I'm not sure if there is a sensible way out. For example, I always wanted to spend some time in Scandinavia. It's not about the money or unvested shares here either. We have some of the best research and development facilities anywhere in Europe here. If I move away, I'll lose everything. I've lost my man; I'd lose everything else."

Christina and I both paused; Juliette's seemed like a plausible story. She seems to have grieved over Levi, to have an impossible job, but she was well-paid, evidenced by the Porsche and the Lake-side Chalet.

"It's not just me, either, I think Amy van der Leiden, Hermann Schmidt and Rolf von Westendorf are all in much the same situation. It sounds corny, but we are trapped by our own success."

"What do we think about the Simon situation?" I asked, "Is he an international man of mystery who has just melted away into the night?"

Juliette mused, "Well, it seems very strange that he would just leave like that, taking all his clothes and everything. Not saying anything to Bérénice or to you, Matt. You seemed to be friendly with him."

"But I get the impression that you didn't like him particularly?" I asked.

"Well, it wasn't a firm opinion. I used to dislike him turning up on our doorstep to go out on the Lake with Levi. Levi seemed to enjoy it though, so what could I say? Simon was far more into yachts than I ever was," answered Juliette.

"But you said you thought he might have a man-crush on Levi?" asked Christina.

Juliette, "Yes, I hope that is the right word. A straight obsession. Like idolizing him. I think many straight men end up having man crushes on Johnny Depp or some Floridian team quarter back. It seemed to be two-way as well, where I think Levi liked that Simon knew so much about yachts."

"But it's not sexual?" I asked - hoping to keep this line of conversation going for a while. Two hot women talking about hot men.

"Oh no," said Juliette, "And, Matt, I can see what you are doing!"

Christina laughed too; they had me busted.

"But enough of this," said Juliette, sipping her cocktail, "Christina, tell us something about life on the road with a band!"

Thursday Research gig

Ever since my first night out with Schmiddi and Rolf, I'd formed a bond with Rolf. I'd been talking about my thoughts on HCI (Human Computer Interfaces) and he tended to agree with what I was saying. In fact, he'd already built a rig in the Lab and was testing some of the ideas that I had.

Our original discussion had happened early in the evening before we both got too drunk to make sense. Hermann, Rolf and I were sitting together in the L'Usine, which was a repurposed factory that transformed into a loud nightclub from the early hours. Early it was a quiet place to go for a drink, and quite close to my Apartment.

Me: "All of these systems seem to rely too much on opening up the skull and then performing brain surgery."

Rolf: "Yes, that was my thought too. No one in their right mind would allow that to be done to them."

Hermann: "But what about using the casque de cerveau

- uh the Brain Helmet - what's it called - oh yes, the ServoCask now renamed as the Cyclone?"

Rolf: "It's okay but lacks fidelity and is really too slow. By the time it has worked out what it has read, processed it, and then created a response, whole seconds have passed."

Hermann: "Yes, I'm more concerned about all the wiring. Imagine an infantryman with that many wires coming out of his head!"

Rolf: "Yes, I think a new piece of combat equipment will be bolt cutters."

Me: "I have to see this thing, but I assume it uses inductance to infer what is happening in the brain?"

Rolf: "That's right. The signals are pinged out like sonar and received back for interpretation and processing."

Me: "So you'd need a HUD - head up display - to be able to read the interpretation of the signals. It's hardly jacked into your own core reflexes then?"

Rolf: "I'm thinking that there has to be another way."

Me: "I've been thinking about that. Using basic anatomy, there's a wiring diagram for all humans. I'm thinking of finding another access point - maybe the tail bone."

Rolf: "That's exactly my thinking. The central nervous system (CNS) is made up of the brain and spinal cord. The grey and the white.

Me "Yes, I was dividing it up into grey matter and white matter. The grey matter, like the brain itself, provides the processing, but the white matter, like that which runs through the spinal column provides the wiring. The local area network if you will."

Rolf (excited): "Exactly. So Like in a local area network, we can put some of the processing elsewhere!"

Me: "Yes, like adding the processing to somewhere with good communication into the brain. The base of the spine springs to mind. The human tail bone finally gets a use."

Hermann: "What! You'd put a processor on the coccyx?! It'd be like those old dinosaurs - what the Stegosaurus!"

Rolf: "No - it's all a myth about two brained dinosaurs. They just had one small walnut sized brain up front."

Me: "Yes but most animals do have a swelling of the spinal cord to control motion, so the idea still - hah hah - 'has legs' as we Brits say!"

Hermann: "encore une fois?"

Rolf, "absolument, trois autres bières, je pense "

Me: "We just ordered three more beers?"

Hermann: "Oh yes, this is thirsty work."

Me: "So an easier siting for an outboard processor - directly wired - might be at the end of the spinal cord?"

Rolf: "You are talking about my specialism now: the interaction between the CNS and the peripheral nervous system (The PNS). "

Me: " It's like having a computer centre - the grey matter - and connections - the white matter,"

Rolf: "Yes. The grey matter in the brain is a major component of the central nervous system comprising neuronal cell bodies, neuropil (dendrites and unmyelinated axons), glial cells (astroglia and oligodendrocytes), and capillaries. This area is 'uninsulated' and so 'sparks' can jump from one part to another, permitting thought and memory."

Me: "It's like the Central Processing Unit and Memory Banks of a computer"

Rolf: "Then there's the white matter in the spinal cord which functions as the "wiring"; primarily to carry information. The white matter is white because of the fatty substance (myelin) that surrounds the nerve fibres. Axonal myelin acts as an electrical insulation. It allows the messages to pass quickly from place to place. "

Me: "The network; thinking about it, even the early biologists knew to label it as the Central Nervous System and the Peripheral Nervous System."

Rolf: "Then consider: Cerebral and spinal white matter do not contain dendrites, which can only be found in grey matter along with neural cell bodies, and shorter axons. White matter modulates the distribution of action potentials, acting as a relay and coordinating communication between different brain regions."

Me: "All of this is making me think we could add a chip into the base of the spinal cord, to tap into the brain signals, like on an Ethernet, we can add a new server anywhere."

Rolf: "And now we are on to the secrets of the white matter: The spinal cord white matter is subdivided into columns. The dorsal columns carry sensory information from mechanoreceptors (cells that respond to mechanical pressure or distortion). The axons of the lateral columns (corticospinal tracts) travel from the cerebral cortex to contact spinal motor neurons. The ventral columns carry sensory pain and temperature information and some motor information.

Me: "The spinal column has to perform several functions and fast enough for the brain to make sense of what is happening. A simple example is typing or playing music. Hand to eye co-ordination must be in real time for this to work."

Rolf: "Yes, long thought to be passive tissue, white matter actively affects how the brain learns and functions. While grey matter is associated with processing and cognition, white matter modulates the distribution of action potentials, acting as a relay and coordinating communication between different brain regions. The brain can adapt to white-matter damage by finding alternative routes that bypass the damaged white-matter areas; therefore, it can maintain good connections between the various areas of grey matter."

Me: "It's like the internet then, finding alternative routes if one gets damaged?"

Rolf: "Remember, we are all Apex predators, so a good line of self-defence is required."

Hermann: "So you two think we can put a socket on someone's butt, instead of in their head?"

Rolf: "I'm sure it can be tastefully done."

Hermann: "I'm not sure which is worse. Brain surgery or butt surgery?"

Rolf: "On that note: *encore une fois?*"

Hermann,"*absolument, trois autres bières!*"

Rolf: "English Man, it will be your turn to ask for the next three beers!"

Rolf Westendorf

Thursday and I was working in Rolf von Westendorf's area of the lab.

It was different from most of the others. Instead of brain cutaways, he had skeletons on display and complicated wiring sheaths displayed all around.

He had been trying to do what telecoms engineers do every day. To trace the wiring from a distribution box to the individual extensions spread around.

Of course, he was working with the human body and the distribution box was the small area at the base of the spine. Instead of looking at the coccyx, he was examining the area above it, which is called the sacrum. It was a much larger surface area and the separation of the two lanes of traffic was more obvious. There were the afferent pathways moving data to the brain and the efferent pathways moving responses to where they were needed, such as hands, legs, feet, toes.

Part of his lab area reminded me of the work I'd seen in

Heather's lab back in Cork. Rolf had his share of rats with back-packs except they looked more precision applied than the ones that Heather had looked after. In fact, there were rats and rabbits which seemed to have gone through similar adaptations., but with the backpacks of circuits connected to - er - their lower regions.

Curiously enough, the animals seemed undeterred by their physical adaptations and I wondered if it was like when a cat gets given an artificial limb, it just accepts it as part of the way things are.

Rolf had some of the mad scientist about him when he was working in the lab. He'd designed an interface from the rats and it could be connected to the Cyclone headgear. It reminded me of my original brain stimulation units that I'd used when a student only about a thousand times better engineered.

I wondered whether Rolf would ask me, and then he did:

"Yes, I've completed the design linkages now for the Cyclone to connect to the rat's efferent signalling. It means we could use the Cyclone to get access to what the rat is thinking and doing. I can't test it on myself though, because if I'm using it then I can't also operate the controls."

Rolf looked at me, waiting to see if I would take the bite.

"Rolf, oh, sure, you can try it on me. I've used brain-stim in the past, so I'm quite used to the idea. But you'll have to tell me - which signal paths work on this test rig?"

"I've included everything we've worked on so far. Afferent and Efferent. Inputs and outputs. I'm just not sure which of the pathways will do anything. Look, I've also wired it to this hand controller. It's like a toy car racing controller. You have to keep it depressed for the signalling links to work. Let go or release the pressure and everything will cut out."

"What about if I go into a muscle spasm and grip the controller tightly?" I asked.

"I've thought of that too, Look:"

He showed me the Scalextric controller and a small gauge: "You need to keep the needle in the green zone for it to work. Too low or too high pressure and the machine will cut out."

I looked at this latest steam-punk addition to the otherwise Swiss precision engineered kit and wondered if I should find a pair of leather and brass goggles to wear.

"Okay," I said, "Let's do this thing!"

Rolf said, "We'll start by seeing how you do with the basic unit. Then I'll add in a couple of stimuli. It's best that I don't tell you anything in case it affects the outcome. I'll record the whole session too, both to video and the readouts from the system logs. Do you mind if Juliette and Hermann watch?"

"Sure, they can help you switch it off too."

Rolf said, "Let me remind you that you are not directly connected to any of the equipment, it is purely through induction of magnetism and light inside to the Cyclone helmet. The rat, on the other hand has direct connection to its spinal cord and onwards to its brain."

Juliette and Hermann arrived. Juliette helped my put the headgear on correctly. And Rolf connected a small plug and socket to the rat's backpack of electronics. The filigree thin wires gracefully crossed into the experiment area. The rat continued as if nothing was happening. It reminded me of one of Heather's favourite rats. One called Montmorency.

"Okay people, experiment time. Are you ready, Matt?"

Showtime

Rolf handed me the toy racing car controller. I pressed the plunger and the dial shot around into the green. To begin with, nothing else happened. The rat continued and I couldn't feel anything different.

Then suddenly, I was over an abyss. My brain had emptied, I had no thoughts - probably the first time that my inner monologue had stopped. I felt drained.

The rat was now motionless. I tried to lift my left leg, and I saw the rat do the same. Then the right. My entire brain was given over to controlling a small rodent.

Then the rat jumped.

I was overcome with a flood of emotion. Food.

The rat could see a small container of food being placed in the experiment area. A big Mac. I felt myself being operated and dragged to where the food was.

The rat ate the corner of the burger. I felt good.

And then another adrenaline rush.

A second rat, larger than the first one, was now in the experiment area and had seen the food. It was going to fight me to secure the food for itself.

Adrenaline.

I gripped the controller harder, ready for the fight. It didn't come. I'd overloaded, and the system had thrown me out. I could feel my own natural thoughts returning and also see Juliette lifting the second rat from the experimental area and placing it in a cage.

"Whoa," I said, "That was intense. My entire worldview was reduced to that of the rat."

"Did you keep your own memory?" asked Rolf.

"No - after I'd adjusted to the headgear - or the headgear had adjusted to me - I felt my whole conscious drawing away. I literally had no thoughts. I was running on reflex.

"When the rat looked still and then I moved it by moving my left and right legs - it used my entire thought capacity.

"But then, when food appeared, the rat could drive itself, but after it had eaten some, I felt good. And then I felt agitated when the second rat appeared, like I was ready to fight for the burger. It got too intense, I pressed the button involuntarily and the system ejected me."

"That was excellent," said Rolf, "We got far more from that than I'd ever expected from a first session."

"Shall we tell Amy?" asked Hermann.

"Maybe not yet," said Rolf, "Let's write it up and submit it with the video and some analysis."

Juliette added, "With so many wires, we should still add some telemetry to the human subject. Heart rate, blood pressure, that type of thing."

I remembered I was still wearing my Apple watch.

"I've recorded some stats because I was wearing my watch," I said - "let me take a look."

I dialled up the history of the last hour on the watch.

"Oh my god! I said, "It's showing a massive beats per minute increase. Not from the start of the experiment, but from when the second rat was introduced. My heart rate lifted to over 220 bpm."

Juliette said, "That's paroxysmal supraventricular tachycardia (PSVT) - about as high as any normal person could go under intense stress. Your heart was trying to pump extra oxygen to all parts of you. Remember a rat's heartbeat is 360-400 bpm. It assumed you could match it."

I felt my legs weaken as Juliette explained.

"Maybe I need to lie down for a minute," I said.

What happens when an algorithm labels you as mentally ill?

Next day, Friday, Rolf had packaged the findings, and we presented them to Amy. She was pleased, but I could see she looked concerned.

"We ran the analytics we captured through Selexor as well," said Hermann, "Selexor thinks Matt is a rodent!"

Everyone laughed. Well, except me.

Juliette looked across towards me and said, "They are pulling your leg,"

Amy spoke, "It illustrates how sluggish the Human Computer Interface is, and also that even the best AI systems today are notoriously prone to misunderstanding meaning and intent."

"We can hook a human to a computer system, but the outcome isn't ideal, far from it. It's slow, has latency, is extremely rough around the edges, creates unpredictable body alerts in the human. For example, the adrenaline and heart rates rise, which could be lethal. If we are using a rodent now as a control, we would need to get much better before we could raise the stakes to a human."

Amy added, "Systems like Createl, have become quite skilled at spitting out data points that seem convincing, even when they're not backed by science. Adding together Createl and Selexor and using its charisma of numbers is troubling. It already gives employers overconfidence when deciding applicants' careers. This experimental work by Rolf and Matt is truly ground-breaking, although troublesome for our unit."

Amy continued, "Even if we can get it to work - the Human to Machine Interface, I am concerned that perceived success at augmenting a person's true intelligence could hide a multitude of sins."

Hermann said, "Joking aside, the Selexor readings are not good. It portrays Matt (when connected to the rat) as mentally ill. This isn't a great advertisement for a Human Machine Interface."

"But isn't that what Kjeld has tasked us to do?" asked Rolf, "To show everyone how great the linkages between our systems are. That we can integrate an amazing, augmented human. After all, Brant wants to signal to the market about how successful they are, and now that Raven has bought Selexor, well - they will consider it to be showtime! Connect Selexor to Createl, plug in the Cyclone and we've got an artificially augmented human."

Hermann spoke, " 'Would that it were so simple,' as my

English friend might say. We all know that the entire chain of components is carrying faults and that it runs too slow to be usable at the moment."

"But will that be enough to deter Brant and Raven?" I asked.

Text from Bigsy

That afternoon I had a text from Bigsy. It was about the gadget we'd found in Simon's apartment. Christina must have passed it to Bigsy.

"Call me," it said, "Gadget time!"

I was in the Lab, but called Bigsy and he was straight onto it, "The lime green gadget from Simon's apartment is a small Wi-Fi receiver, which hooks onto the Wi-Fi in Simon's apartment. It's made in China and has a built-in name of Клиент_Sacha, and a config language of Russian. As a matter of fact, it looks as if it was talking to the unit you sent me a picture of, from your own apartment."

"How can you tell?"

"There's some small writing on the other device - the one in your room. It's the MAC address. A media access control address is a unique identifier assigned to the gadget. Client_Sacha appears to link to the MAC address in your room. And your unit? It was a receiver for audio and video feeds. A basic room bug, if you will, listening to wireless devices in the room. They are easy to obtain nowadays, for a few hundred dollars."

"So my room was bugged, and it was sending its information to Simon?" I asked.

"Yes; and it was using the gear you can buy in an online spy shop. Nothing terribly sophisticated."

"What about now? I've re-run the scanner and there doesn't seem to be anything else running."

"If you used the gizmo I sent you as the scanner then you should be in the clear," answered Bigsy, Just unplug the flat black unit on the back of your television."

"Okay, that's good to know. I can make calls from my Apartment and invite friends around!"

"Oh yes, knock yourself out!"

At last, I could invite Juliette to the Apartment.

"Something else?" said Bigsy, "Jake has been doing some digging...Hang on, I'll put him on."

"Hello Matt, or should I say 'bon nuit?' "

"Hello Jake, Bigsy tells me you are on to something?"

"Yes, I followed up on that yachting story that Simon had told to Juliette. I think she has misunderstood a part of it."

"What's that then?"

"About the Barents Sea. I don't think it is the actual sea, I think it was a reference to a particular yacht. My buddy Jean Sauveterre, who I know from when I was on a magazine called 'Street' was mad about big boats."

"We both used to interview young guns who spent inordinate sums on cars and other toys. We called the series 'fast boys'. I did cars, and he did superyachts and sometimes we shared doing the planes. The men who owned this stuff were super-rich and it was often from dubious sources. Imagine being 26 and owning a McLaren supercar? Or showing off the lifestyle on a party boat in Nice? These guys did it all the time.

"Anyway, I contacted Jean and told him about the yacht and the Barents Sea. He laughed and said it must be a bulk carrier!"

Jake continued, "Jean said he thought the Barents Sea was probably the name of the vessel, and sure enough he soon found one, registered in Malta and operating in the Mediterranean. He said it was a classic party palace, Italian-built, 43 metres long, took a crew of 8 and could host 12 people in a Master state-room, VIP suite and some doubles. It had the jacuzzi and pool and a couple of 10m and 5 metre tenders. He reckoned they would rent it for around US $140,000 per week."

"So, a super yacht, not something you'd drive around Lake Leman?"

"Yes, although super yachts only have to be bigger than a tennis court to qualify for the name. Greater than 24 metres is all it takes, or so my friend Jean has explained to me on numerous occasions. Then there's the mega yachts. Think of Sir Philip Green's Lionheart, which is around double the length of Barents Sea, has a crew of 40 and a helipad. He sold it after the trouble about the company pensions. But don't worry; he has two others."

"Erm, isn't this getting to be excessive?"

Jake continued, "Jean says the industry keeps many people in well-paid employment. But newbies must decide early on whether they can sit by and watch their billionaire bosses spend vast amounts enjoying themselves, while appearing not to care about those around them. Owners expect the best in the world. They want to go wherever, whenever, and demand the highest standards without delay. Money is not a problem for them."

"Okay, so we've another perspective on Simon's yachting now."

"Yes, but that's not all. I told Christina about it and she recognised the setup immediately. I guess you will realise that Christina has some unusual perspectives on things?"

"Yes, she seemed scarily professional investigating Simon's disappearance around here, and the way she quietly cross-examined Juliette. Is she a cop or something?"

"Well, Christina recognised the style of operation that Simon appeared to use. Apparently, the Russians will sometimes use the Mediterranean as a starting point for their security and surveillance for their own. You know, the oligarchs who swan around Monaco and the Greek Islands?"

"I can imagine," I said.

Jake continued, "Well, if you want to borrow a super yacht to float around in sunny places, you can visit the Med, or the Tax Islands, or either coast of America. The Russians like the Med and can show off to one another about their super-toys. They'll also hire crew who can speak their language and are usually interested in crew who can provide security as a bonus.

"Christina said it was therefore natural for the Kremlin to want to provide some useful agents around the Med, who could operate yachts, speak Russian and also be handy with an A.K."

"A.K?" I queried.

"AK47 - Kalashnikov - a submachine gun," explained Jake.

"Duh - I see - this has just got very dark."

"Christina said the moves were fairly predictable GRU tactics. Embed an agent - that's Simon - he befriends the mark - that's Levi - and the next thing you know Levi is floating on a lake as a result of a boating accident. After a suitable pause to deflect suspicion, Simon disappears, presumably to re-surface in Cannes or Nice or somewhere with plenty of yachts and Russians."

"Wow - that's...that's ruthless."

"Yes, it implies that the stakes for this work that Brant is doing must be sky-high."

"Have you told Amanda Miller about this?"

"Oh yes, and she is about to send in some cavalry to provide additional protection. People who have experience of dealing with Raven and Brant- but for that, I'd better let you speak to her yourself."

Prising the lid off

In the evening, back at my apartment, I wondered if it was really safe to make a call to Amanda. I'd pulled the small black unit from behind the television and unplugged its connectors. It lay on the table.

Then I had an idea. Finally! A use for my Swiss Army knife. I carefully selected its large blade and prised open the black box. Inside, to my perverse delight, I saw it had two rechargeable batteries. Now I could use the 3mm screwdriver blade to prise them out! With a plink, they fell to the floor. I assumed I had now disabled the unit. It was damned sneaky of whoever designed it to make it self-powered. I had no-idea how long the batteries would last, but I thought it was a good question for Bigsy at some point.

I could see inside the unit there was a small identification label. It said it was a Xiaomi AC2350. I decided to hit the internet. Sure enough, the same make as the small lime green Wi-Fi receiver, made in China, and approved for use in India.

This one had a few extra interesting things about it. The first was it showed the picture with seven antennae, notably missing from this installation. It also described it as an IoT (internet of Things) router, so it was purpose built for surveillance. And the originals didn't appear to have batteries inside. This was a special adaptation. The batteries didn't give anything away though. Generic AA rechargeables from Amazon, soldered together and slimly taped into the box.

The main thing was that the spy unit was now dead. I shuddered at the thought that there could still be cameras and microphones scattered around the apartment, but they would now be reporting their findings to nowhere.

Then, I called Amanda. She sounded as if she was walking, and I guessed she was on a headset.

"Hi, Matt, I hear you have been busy! Jake, Christina and Bigsy have been keeping me informed. Bigsy tells me your Apartment was bugged and you suspect Simon Gray? And Jake says Simon Gray was a possible mole? I am checking the way that I was handed Simon right now. I must tread carefully though; in case I am about to kick over a hornets' nest."

"I can see that, if you were given Simon as a trusted member of MI6, but yet he is an embedded foreign agent."

"Correct, and it means that Jim Cavendish (whom I trust implicitly) and I have both had the wool pulled over our eyes. Jim gave me the name of Simon, but someone will have had to provide it to him in the first place. Look, can we continue with this briefing tomorrow, Saturday, during the day? I'm about to get on a plane. Tomorrow will also give me a chance to round up a few people. We can run it as a full conference call, and we'll know that everyone is on the same page?"

"Sure, I'm in my apartment and think I've disabled all of the listening devices - at least according to the gadget Bigsy sent across."

"Okay then. I'll send you a link for our call tomorrow."

With a click, Amanda was gone. She was one busy lady.

Saturday VTC

Saturday morning and I realised that, once again, I wouldn't get a lay-in. Amanda had set the video call for 09:30UK - which for me, luckily, was 10:30, but by the time I was presentable and functional meant I couldn't have a long lay in watching a box set.

I dialled in promptly and was greeted by Amanda, the Triangle folk, Chuck Manners and Grace. A full house.

"Hello Matt, and welcome to the call. Let's summarise what we think we know:"

I noticed Amanda say 'think we know' implying some of it was still under review.

She began, "We know Brant is making a Human to Computer interface. That there are various components of it which don't work properly. They want to sell it. We suspect it will go to the highest bidder. Brant's and Raven's game is to raise their own share price on the strength of a large sale.

"Levi, we think, knew that the AI component called Createl which he had invented for crop evaluation couldn't be transformed into something that could do image recognition of faces on a large scale. He would have blown the whistle on it, except he died in a boating accident."

"The AI component which Levi's Createl linked to is called Selexor but it is also flawed. It is supposed to hire recruitment candidates, but is driven by manually constructed rules, like 'no one over 1.8 metres high' and so on."

"And then the headgear, Cyclone, looks impressive, but only works at a very slow speed, and requires many wires attached to it.

"Put the chain together. Selexor (dubious) plus Createl (flawed) plus Cyclone (Sluggish) and the current invention doesn't work."

"They realise that in my Research Team, "I chipped in, "Everyone up to Amy van der Leiden, our team leader knows it."

"Yes, then we can add in the other events. Simon Gray, our MI5 operative, is embedded in Brant. He appears to speak Russian and to have been bugging at least Matt's apartment. He befriends Levi through a common interest in yachting but then soon after Levi dies in a boating accident, Simon completely disappears. We are still checking how we at MI5 came to be provided with Simon for this assignment, and to track where he has gone to."

"Have I missed anything?"

"Only that the listening bugs used in my apartment were Chinese, but had been programmed by someone using Russian names," I answered.

"Do we think we know what Brant wants to sell?" asked Clare, "It sounded like it had a military purpose."

"Correct," said Amanda, " Brant is well known for its military subcontracting. It will go into a war zone after the conflict ends and rebuild both the military complexes and also the towns that have been destroyed."

"It sounds a little like the moves that Halliburton made after the Iraq war," said Chuck, "I remember Vice President Dick Cheney's former company, worked alongside US troops in Kuwait and Turkey under a package deal worth close to a billion dollars. They built tent cities and provided logistical support for the war in Iraq."

Jake added, "It was all packaged up under the banner of the 'war on terrorism,' I seem to remember,"

Clare added, "The reconstruction gravy train, and one that Brant has successfully hooked its engines to in more recent times,"

Chuck added, "It was commercially ruthless really, like Kellogg, Brown and Root, which operated as a subsidiary of Halliburton and secured that 10-year deal known as the Logistics Civil Augmentation Program (LOGCAP). The contract provided a "cost-plus-award-fee, indefinite-delivery/indefinite-quantity service". It meant that the federal government has an open-ended mandate and budget to send Kellogg, Brown and Root anywhere in the world to run military operations for a profit."

Amanda added, "Not to mention the later Restart Iraqi Oil (RIO) contracts. Yes, now that's where suspicion creeps in. When we've had dealings with Raven - who wholly own Brant, we have had reason to see them associated with Russian influence."

Christina chipped in, "Yes, but Russia nowadays operates asymmetrically. It's been one of VoVa's big ideas. Cut the cost of winning by using cheaper methods than the perceived enemy. It's like stacking the votes in an election or taking down Critical National Infrastructure with a few viruses."

"Vova?" I asked.

A chorus of "Vladimir Putins" greeted me. I was the only uncool cat who didn't know his nickname.

"We had a name for the KBR moves at the Dominion Academy; it was a case study," said Christina, 'Wartime and Disaster Profiteering' - about how Dick Cheney could use his old buddies to turn a handsome profit and share in its proceeds. I remember the case study also looked at other countries like Turkey."

"I remember spending some time in Turkey," said Chuck, "The Americans needed somewhere to station their Air Force F-15 Strike Eagles and F-16 Fighting Falcons. They were monitoring the no-fly zone above the 36th parallel in Iraq. The jet pilots were catered and housed at the Incirlik military base seven miles outside the city by a company named Vinnell, Brown and Root (VBR), a joint venture between Brown and Root and Vinnell corporation of Fairfax, Virginia, under a contract which also includes two more military sites in Ankara and Izmir - everything was funded under 'war of terrorism' terms."

"Good," said Amanda, "We've got plenty of context now, so Raven and Brant operate under terms like those created way back around the time of the Iran conflict."

"Christina, do you remember when we went to Raven's Headquarters in Austin, Texas?" asked Jake," To see that guy who had supposedly been in a fire-fight in Iraq?"

"Oh yes, Kevin Dubner!" she said, "He spun us a lot of American military jargon about 'death blossom' gunfire and NSTV - that's non-standard tactical vehicles - modified civilian gear. His story was almost too word perfect - that is right up to when he got his 939s and 977s mixed up. He didn't know his front-line troop carriers from his container haulers."

"And the point of this is?" I asked.

"It shows that Brant wanted to hide something back then and right now. They appeared to be provoking small fire-fights to prolong the war, keeping US troops on the ground and their all-important logistical supplies open."

"Leading to more contract extensions," said Clare.

"And asymmetrically achieved as well - low cost, high gain," said Christina.

Grace cut in, "Well, I'm not sure how this fits in, but it looks as if Brant has also been making moves toward the Chinese. Some of this will sound like Robocop, but the Chinese are interested in developing soldiers with 'enhanced capabilities.'

Chuck added, "I've heard about this too: China is conducting tests on its army hoping to create biologically enhanced soldiers, according to the Pentagon.

"John Ratcliffe, who was Donald Trump's director of national intelligence, made the claims in a newspaper editorial, where he warned China poses the greatest threat to America today."

"I have the article, from the Wall Street Journal," said Grace, "It was also in the UK Guardian."

"Although it was back in the time of Trump," said Jake, "So try not to inhale too much."

Chuck said, "Yes, the reports stated Beijing intends to dominate the US and the rest of the planet economically, militarily and technologically. Many of China's major public initiatives and prominent companies offer a layer of camouflage to the activities of the Chinese Communist Party."

Jake asked sceptically, "Or was this Trump trying to Make America Great Again and to stop the influx of Chinese product to the American shores?"

Chuck continued, "It sounds like the things that have been happening in Brant's labs in Switzerland. Our US intelligence showed that China has even conducted human testing on members of the People's Liberation Army hoping to develop soldiers with biologically enhanced capabilities."

"Yes, that sounds like the tests that we are conducting, although it implies that the Chinese are further advanced," I said.

Grace, "At least in their propaganda,"

Clare had been busy, she flashed up a screen with a summary:

- Selexor (dubious)
- Createl (flawed)
- Cyclone (sluggish)
- Levi - Sceptical of the leap from crops to people
- Simon Gray - killed Levi?
- Simon Gray - evidence of surveillance of Matt + others?
- America - suspects Brant of wartime and disaster profiteering
- Raven / Brant want to sell the alternative solution to military
- Suspect Russian influence through Simon Gray + Brant
- Chinese experimenting with similar weapons to Brant

"Clare - you are always so good at the summaries, " said

Grace.

"Thank you. My marketing background!" replied Clare.

"Okay, so what will we do next?" asked Amanda, "We need to put a stop to this. I keep saying this around events involving Raven and Brant."

"Well, we could try to locate Simon Gray," said Grace, "There must be movement record for him. Passport, cell phone, CCTV, something?"

"Good point," I said, He didn't have a car so would have need a cab or something to get from the apartment on Rue de la Confédération."

"And with luck we can find if there are any CCTV around that area," said Grace.

"Another factor is the fragility of the HCI design," I stated.

"No, we should leave that as a lure for them to keep on working," said Amanda, "We can try to flush out who is really keen to see it get sold."

Amanda answered, "Yes. And we need to consider the other members of your team. Are they all what they purport to be? - Juliette Häberli, Rolf von Westendorf, Hermann Schmidt, Amy van der Leiden?"

And then Chuck asked, "What about in your Apartment? Is there anyone else suspicious? You said Simon had a girlfriend? What was her name? Bérénice Charbonnier?"

Now I had the research to do, a separate investigation to conduct, and Chuck and Amanda were urging me to be careful who I should trust.

"What about if I have a party?" I said, suddenly inspired, "Nothing too fancy, but I could get everyone around to mine, a bit like the party they invited me to, over at Bradley Floyd's. I could use it as a chance to say thank you to everyone for making me feel so at home, but I could also use it to check people out?"

"That's actually a good idea," said Amanda, "But with just one of you, Matt, it will be difficult to get around to everyone. I was thinking of doing this anyway, but I think it could coincide with your party?"

"What's that?" I asked.

"I was thinking of sending Chuck and Christina over to provide you with some support, in case things got difficult," replied Amanda, "Neither of them will be known either, and I don't have to worry about diplomatic passes."

Chuck answered, "That's fine with me. I haven't been to Geneva for years but will look forward to it!"

Christina smiled, "Yes, it will be good to meet up with the folk I met a few days ago. Now I know more about the situation I can have some better questions too!"

Jake added, "I wonder if I should go as well. I'm never one to turn down a good party, and I'm really quite good at working the room - my journalist training!"

"Okay - next weekend then, here at mine!"

"I know a good hotel too, that's walkable from Matt's," said Christina.

Party at mine

Sunday, I printed some individual invitations and took them around to the other apartments.

I made sure to invite Bradley and Jennifer, Bérénice and even Father Oscar, via Bradley.

They all mentioned that the date I'd picked was around Dèsalpe - when the cows come off the mountains, to overwinter out of the snow. They told me that the nearby town of St Cergue also celebrated the colourful, traditional Festival Dèsalpe.

By luck I'd picked the end of the Alpine summer pasture season and the day after my party only 35 km away there would be a colourfully decorated parade of cows, goats and pigs through the valley to the sound of bells and alphorns. Swiss food stands, music and a very festive atmosphere would make a great day to extend the party.

Oscar also told me he knew some musicians who could play Swiss folk music, through his church. This way I could create a Dèsalpe mood, complete with entertainment. I also asked Aude if she would like to drop in, and when I told her the date and occasion, she said, "How could I refuse, it would be lovely."

Sunday evening, I also sent a few emails to the people in the Lab. I mentioned the party and went so far as to call it a Dèsalpe party.

"Oh yes, Comical to have an Englishman throwing a Dèsalpe party!" replied Rolf, with smiley face. Hermann confirmed he'd be along, "I'll bring some Schladere," he said. "Proper Swiss Schnapps."

"I'll get some cheese then, " I said replied, "And try to arrange for a few bottles of English beer."

"I'll help you with this," emailed Juliette, "We can get the British beer at the Intermarche St Genis and I'll bring them around in my car!"

This was great. The idea of the party seemed to be taking off nicely. I thought I'd keep the theme a surprise for the people coming in from the UK.

This is awkward - notes about Brant

Monday Morning.

Juliette came into the Lab, somewhat distraught to have seen an article in the Sunday papers. It was in the La Liberté de Geneva and was a syndicated article from a British tabloid.

It seemed to be a direct critique of the work that Brant was doing in the Laboratory, and in particular the modern warfare element. She said it implied that machines were taking over warfare and it talked about 'killer robots'.

Juliette spoke to Rolf, Hermann and me.

"This wasn't why I joined the Laboratory," she said, "I wanted to promote peaceful uses for the technology. To support people with healthcare issues, not to provide new ways to give them the issues in the first place."

Juliette read a section of the article, "The US Air Force has predicted a future in which SWAT teams will send mechanical insects equipped with video cameras to creep inside a building during a hostage standoff".

Juliette continued, "A microsystems collaborative from Berkeley has already released Octoroach, which is an extremely small robot with a camera and radio transmitter that can cover up to 100 metres on the ground. There's another one with two legs and a 'copter called BOLT which can fly through rubble.

"These weapons are 'biomimetic,' or nature-imitating, and are on the technological horizon."

Schmiddi interrupted, "Yes, but expense is the chief impediment to a great power experimenting with such potentially destructive machines."

Rolf added. "Unless the labs run software modelling to allow virtual battle-tested simulations to inspire their future military investments."

Juliette said, "Some of the world's largest militaries seem to be creeping toward developing such weapons. By pursuing a logic of deterrence, they fear being crushed by a rivals' AI if they can't unleash an equally potent force. Another arms race. Boys and their toys against other boys' toys."

Hermann added, "Maybe our Lab is too literal. Trying to provide an AI link to a human. In other words, to make augmented humans look like and behave like real humans?"

Amy entered the Lab and our discussion, "I could hear you talking, Yes, the key to solving such an intractable arms race may not be in global treaties. More in a cautionary rethinking of the uses for martial AI. The deployment of military-grade force within countries such as the US and China is a stark warning to their citizens: whatever technologies of control and destruction you allow your government to buy for use abroad now may well be used against you in the future."

"But isn't this disloyal to Brant?" I asked, "Our whole research programme is aimed at HCI - Human Computer Interface. If we are just going to make intelligent, gun toting wheelbarrows, it becomes a whole different game."

Amy added, "Brant is all about the money. Every time I see Kjeld Nikolajsen, he asks me how much progress we have made. He doesn't want to hear about our troubles, our challenges. He only wants to hear when we are getting closer to his goal of monetising everything. Kjeld doesn't really care whether it is a helmet with a HUD - head up display - a communications device between a human and a computer, or autonomous gun. He just wants to know that we can productise and sell something.

She continued, "Brant knows that weaponry has always been big business, and an AI arms race promises profits to the tech-savvy and politically well-connected. Even better to quartermaster and support the weapons - which is Brant's service offering."

"But can we counsel against it?" asked Juliette.

Amy continued, "Counselling against arms races may seem utterly unrealistic. After all, nations are pouring massive resources into military applications of AI, but it is hidden away as Top Secret."

She added, "Military and surveillance AI is not used only, or even primarily, on foreign enemies. It has been repurposed to identify and fight enemies within. US Homeland Security forces have quietly turned antiterror tools against criminals, insurance frauds and even protesters.

Rolf spoke, "Yes, I've seen those robot devices with caterpillar tracks that look like blue-painted mine detector machines, and were used to police coronavirus lockdowns in Shenzhen, China."

Amy added, "Brant knows that once deployed in distant battles and occupations, military methods tend to find a way back to the home front. They are first deployed against unpopular or relatively powerless minorities, and then spread to other groups."

Juliette spoke, "We've seen the US Department of Homeland Security officials gift local police departments with tanks and armour. Sheriffs will be even more enthusiastic about AI-driven targeting and threat assessment. But it is important to remember that there are many ways to solve social problems. Not all require constant surveillance coupled with the mechanised threat of force."

Rolf spoke, "Indeed, these may be the least effective ways

of ensuring security, either nationally or internationally. Drones have enabled the US to maintain a presence in various occupied zones for far longer than an army would have persisted. The constant presence of a robotic watchman, capable of alerting soldiers to any threatening behaviour, smacks of oppression."

Herman added, "Yes, American defence forces may insist that threats from parts of Iraq and Pakistan are menacing enough to justify constant watchfulness, but they ignore the ways such authoritarian actions can provoke the very anger it is meant to quell."

Juliette cut back in, still holding her copy of the Sunday paper, "It talks in here about how the military-industrial complex is speeding us toward the development of drone swarms that operate independently of humans, because only machines will be fast enough to anticipate the enemy's counter-strategies.

She added, "But this is fed in by Brant's PR and becomes a self-fulfilling prophecy, tending to spur an enemy's development of the very technology that supposedly justifies militarisation of algorithms. I might be on my 'high horse' but to break out of this self-destructive loop, we need to question the entire discourse of imparting ethics to military robots. Rather than marginal improvements of a path to competition in war-fighting ability, we need a different path – to cooperation and peace, however fragile and difficult its achievement may be."

I was impressed with the whole discussion. The team had decided that what we were being asked to do by

Brant was now both flawed and unethical.

Wednesday afternoon

Wednesday in the Lab. Still no news about Simon Gray.

I'd been working with Rolf again, and our experiments with the Cyclone had run into a dead end. We'd been looking into the linkages between Selexor and Createl to see if we could find what was causing the immense delays in the system.

We'd even called on Hermann to help us with some programming of test cases, to see whether we could generate responses from the Createl system. It was simple enough, instead of flying the Createl over a field of corn, to look for bugs, we were flying it over a field of faces to see whether it could pick out the unusual ones. We made it very simple, with around thousand faces from a football stadium and then one face which had mouse ears and a silly hat. We wanted to see whether it could discriminate sufficiently to 'find the mouse.'

We ran it ten times and it didn't ever find the mouse-face. We even boosted the contrast in that area of the stadium photos, but still to no avail.

Hermann had looked through the original software developed by Levi. "It's multi-dimensional. It doesn't just work on one level - It uses heat mapping, imaging, chemical analysis, 3D stereoscopy, timeseries, and statistics to divine the state of a field. No wonder it won't work when it is looking at humans. It doesn't have the bandwidth."

Rolf spoke, "But if we look at some of the Chinese Face ++ software, it seems to be able to handle a dense crowd? What's the difference?"

Hermann again, "It's the different use cases. The Chinese will put cameras all over, say, Tiananmen Square and then track people through the area. They can allocate a `face_token` to the value they've interpreted. Then they can use the `face_token` in image matching later. 'Was Bobby in the Square?' or 'Look for Bobby' Both of this use cases are different from 'Tell us how many people', and 'describe how many types of person' - the last one is a particular AI case where a human can spot something unusual quicker that a machine."

Rolf, "The Chinese must be pretty confident too; they have published the interfaces and even some of the Java Script Object Notation that is achieved."

Hermann replied, "Yes, and they've come up with a second system - akin to our Selexor, which they have called Brain ++ It's a DeepLearning system."

"Do you think its original work?" I asked.

"What do you mean: do we think they've borrowed it? I seriously doubt it, but I won't be surprised if it has been hyped in the same way that Brant's PR hypes its own systems.

Bait

Wednesday evening and Amanda had called for an update from everyone.

Grace had a very simple, if somewhat expensive idea to trace Simon Gray. She arranged for his current bank account to be credited with a large sum. She assumed his plan was to ignore the account, but that by putting a large sum in to it, it would tempt him to break cover to transfer the money somewhere else.

Incredibly, after three days, it worked. But the surprising part was that the transfer had been conducted not from the Mediterranean, but instead it was from the USA.

"It shows the bank transfer run on a PC laptop from the Marriott Residence Inn in Glendale, Arizona."

"What?" asked Clare he's out in the Arizona desert?"

"Not really," said Chuck, "That area isn't far from where we all travelled when we were looking into that business over at Scottsdale."

Bigsy had flipped it upon a map. Sure enough, Scottsdale and Phoenix were close by.

"Still. What on earth would Simon Gray be doing in that area?" asked Clare.

"It isn't that far from where I used to do missile testing with the USAF," answered Chuck, "but, I guess they have moved their testing along to other devices now - Maybe those include robotics?"

Bigsy added, I see the Luke Air Force Base is also close. Does that ring any bells with anyone?"

Chuck started, "Luke? They set up a huge training facility there, for F-16s at the time. Except it wasn't with real planes. They used simulators, out on the tarmac, under a canvas roof. It's the main place where fighter pilots get trained for combat duties in the US."

"Like these simulators?" asked Bigsy, flipping up a picture of a vast area of what looked like staggered rectangular tents.

"Most of those are probably the individual plane hangars!" said Chuck, "They have smartened up the facility since I was there."

Bigsy zoomed in, "Oh yes, I can make out a plane in that hangar!" he said, "And another one parked outside!"

"The one in the hangar is an F35A. See the twin tails?" said Chuck, "The other one is an F-16 Falcon - see it's only got one tail. The F-16s are gradually being replaced by the flying computer that is the F35A II Lightning."

Amanda cut in, "Okay, so Simon appears to have gone to a hotel close to a USAF Air Force Base. Why?"

Christina spoke, "He seems to know a lot about boats, maybe he also knows about planes? - If he had been 'trained' by a foreign power, then this would not be so unusual."

"What are you saying, Christina," asked Amanda, "Do you think he has been to 'spy school?' "

Christina answered, "Well think about it. He's a flaky back story about living in Hampshire, England, yet doesn't know his way around London - barely 50 miles away. He has Russian prison tattoos on his shoulders. He had a collection of bugging equipment to check up on Matt. He knew how to crew a large yacht, at sea in the Mediterranean. A close friend of his mysteriously lost their life on a yacht on Lac Leman. Simon disappeared efficiently, but then re-appears in the USA, at a location close to a US Air Force Base. - You've heard of that razor test - the simplest answer is probably the right one."

I was impressed that Chuck and Christina seemed to have most angles covered.

"Well, I think we should detain him immediately, before he gets away again. It will set off a few alarm bells, but at least we can question him," said Amanda.

"Grace, do we have any direct contacts with Luke Air

Force Base?" asked Amanda, "Only getting the detention of Simon Gray via diplomatic channels is going to take too long."

"I think I still know some of the base," said Chuck, "Remember I was at Kirtland and White Sands. We used to hitch rides on planes and choppers from Luke AFB all the time. And I served in a war zone with Major Ben Randall - who was up-and-coming at the base. He was a highflyer - if you pardon the pun, and I guess he is still there."

Bigsy looked, "If you mean Vice Wing Commander Ben Randall, then he is certainly there! It looks as if he is deputy to the big boss too!"

Chuck replied, "I knew Ben would do great things. I know him from our time together on the Iraqi oilfields. It was a nightmare, and we spent some time healing our mental injuries in the Green Zone. As a matter of fact, I had to help him out of a burning building once. It was a result of a terrorist bomb. He's a great guy. Grace I'll bet you have a number for him?"

Grace replied, "I certainly do. It is a base land line actually, direct to the command centre."

"Okay - I'll call him - right now in the middle of his night. We don't want Mr Gray to get away."

Amanda continued, "What else have we got?"

"I've arranged the party for this weekend," I said, We'll have most of the people connected with this along. You are all welcome, " I added.

"I thought I might tag along," said Bigsy, "I can check out your apartment for other devices at the same time."

"How about you, Clare?" I asked.

"No - Thanks - but I'm already booked this weekend. Otherwise, I'd love to."

"Jake?"

"I can't I'm afraid, I've got a stag do booked with a bunch of ruffians. Not something I can duck. As a matter of fact, we are off to Amsterdam."

"Okay, well if anyone changes their mind..." I thought I'd made my invitation as extensive as possible.

Chuck's email

Chuck emailed us all later that evening.

The USAF had located Simon Gray, staying under the name Sacha Kravets in the Marriott. He'd been quietly apprehended and was now being politely held on the Luke Air Force Base.

They had sent a photograph and it certainly looked like Simon, although the name was completely off, and he was said to speak English with an American accent.

He also held a US Passport, which Grace emailed to say checked out and was for a taxi driver living in Brooklyn. The taxi medallion checked out too, so 'Sacha' had a realistic back-story although it wasn't clear what he was doing in Arizona.

Amanda challenged why Sacha/Simon was now based in Brooklyn, and Christina had the best explanation. "He has finished his previous assignment. They have burned his old identity and given him a new one for the next piece of work. It is the typical working style of a hitman," she added.

"His biggest mistake was to be on an American passport," said Amanda, "If he'd been a foreign national, then he could have used an embassy to remove him from the situation. As an American we have him pinned down."

"But be careful, " said Christina, "If anyone else knows where he is, or that he has been detained, then he could be in great danger."

"They have to get into a high security Air Force Base first," said Chuck.

Kyle Adler

Thursday morning, on the way to the Lab, I received a couple of texts. The first was from Kyle Adler: "See you in Geneva later tonight."

Great. Kyle was able to get away from his assignments and would be joining me in Geneva.

Then a text from Christina.

"We're on the way, staying in same GVA hotel,"

Excellent - They would all be in Geneva early, which would give us a chance to plan anything extra.

I texted Christina for us to meet in the bar in the evening. I'd suggest Kyle join us at the same hotel and then we could all synchronise together.

In the lab, Rolf and Hermann had been inspecting Levi's code for Createl.

It was Hermann who said it first, "It's as if there is a

whole extra layer of software in the Createl design, but I can't work out what it is doing."

Rolf added, "Yes we'd assumed that the base code of Createl was good and pretty much the way that Levy intended it, but the more we drill into it, the stranger it looks."

Hermann said, "I think Levi built the system with a protective layer around it. It seems to need a key to unlock it and then it will run smoother and many times faster."

"You mean it's like a 'diagnostic layer?' I asked, to help iron out bugs and to prevent viruses?

"That's what we thought first, but I think it is something more subtle. Something which downgrades the software unless a security token is served to it," answered Hermann.

"I just heard today that my good friend Kyle will be in town for the next few days. He is a security whizz and may be able to shed some light on this."

"But he won't be cleared to come on property, let alone to see any of the secret stuff," said Rolf.

"I'm going to ask Amy to clear it. I'm going to tell her it should help us," I said, "with Amy's agreement I could bring him in tomorrow."

"With my agreement to what?" asked Amy, who had just walked in.

"My friend Kyle Adler is a security whizz and he'll be in town tomorrow. I was going to suggest he take a look at this extra layer that Rolf has discovered in the Createl software, to try to understand what it provides."

"Kyle Adler?" I saw him present at an event in Israel,

"He's quite a guru on security, isn't he!" said Amy.

"Oh, you know him?" I asked.

"More importantly, you know him, Matt. Sure I'd be pleased for him to take a look at the code that we've been puzzling over - so long as he signs Brant's NDAs."

I knew Kyle would sign the non-disclosure agreement with Brant.

F-16 Advanced Concept Ejection Seat – Mode 1 Deployment

Thursday evening, I'd arranged with Christina to have cocktails in le Bar des Bergues at the Four Seasons.

We'd start there because everyone was staying there. I decided to book a table for a little later in the evening at Bistrot du Boeuf Rouge, only a short walk along Quai du Mont-Blanc and then left along Rue Docteur Alfred-Vincent.

I thought it would be a great way for me to show off my local knowledge of Geneva, and the destination also served fantastic local dishes. Anyone who wanted to leave early would only have about a seven-minute walk back from the restaurant to the hotel.

I arrived around ten minutes early, and Christina was already in the bar, with Chuck Manners, Bigsy and, to my surprise, Clare.

"Hey! Matt!" called over Chuck as we entered the bar.

We all shook hands and did variations on the 'bise' with one another. A waiter looked over and I recognised him from the evening I'd spent here with Christina.

"Bonsoir," he said, "Would you like 'your usual?' " he winked, and I said, "Certainement"

"Oh - showing off already with your new-found French, I see," said Chuck.

"Clare. I thought you said you couldn't come along?"

"I changed my plans. Who wouldn't want to go to a party in Switzerland over a long weekend!"

"My friend from the UK is coming along as well. His name is Kyle - Kyle Adler, although he's as English they come. He's a bit of a security expert actually, and I hope will be handy around the Lab."

Chuck's voice softened, "The others know this already, but I've heard from my buddy Vice Wing Commander Ben Randall. It seems that Simon was secured on the Luke Air Force Base in Arizona, when two Federal Agents came to collect him. They took him from the secure area and planned to move him into an armoured van. Somehow, he escaped with an airman, disguised himself as a pilot and then he and the other airman stole a fighter plane. "

"It sounds like something from an action movie!" I said, incredulous.

Chuck pulled a face that made me think this story

wouldn't end well. "They successfully took off in a two-seater F-16D Viper and were making a low altitude escape when it looked as if the plane had gone out of control. The co-pilot ejected, but Simon was left in the plane, travelling at close to Mach One, when it crashed to the desert floor around Canyon de Chilly, which was about 300 miles - that's less than twenty minutes flying time - from Luke AFB."

"He crashed into the desert, at near Mach One!" I shook my head. Then Kyle arrived, oblivious to Chuck's story telling.

"Kyle!" I called.

"Mate!" he called back and walked over. Another round of handshakes and greetings and Kyle wasted no time in ordering a Namur Express - "Belgian bier brewed by the Swiss," he explained, "And sorry Chuck - I think I interrupted you in the middle of a story!"

Chuck continued, "Yes - just as the plane crashed into the desert floor! Well, the co-pilot who had ejected wasn't found and Simon's ejector seat hadn't been fired, despite F-16 ejector seat's record of successful deployment as late as one second before a crash in other situations."

I reeled at this news. Simon had seemed to be a pleasant enough guy and been helpful to me when I was getting established. We'd sat together on the bus to the campus, and I'd regarded him as one of my earliest friends in Geneva.

"Plane crash?" asked Kyle, looking shocked.

"Yes, someone Matt knew, and we'd all met" said Christina. Kyle grimaced.

Chuck continued, whilst the others sat in silence. "The same lack of traceability goes for the two Federal Agents, who also disappeared. We can't tell whether it was a legitimate escape attempt or a cold-blooded execution, covering their tracks. Nor can we even tell whether the Federal Agents were imposters."

Chuck added, "To be honest, I think this smacks of a well-engineered put-up job. Luke Base handles both training but also fighter planes. There are B-61s on site in Luke, so to fly out an F-16 is almost ridiculously lax of their security.

"What's a B-61?" I asked.

Kyle answered, "It's a nuclear bomb. A compact 3.5 metre long by 0.3 metre diameter bomb - one that can be attached to even a fighter plane like an F-16. They are dial-a-yield bombs that go from 0.5 to 340 kilotons. That's about ten times the power of the ones used in World War II."

Chuck nodded, "You're good. You know your armaments."

Kyle answered, "Only because I protect them with advanced security. It sounds as if the Luke USAF base needs my help. There's over 540 of those B-61s out in the wild, enough to bomb the shit out of the earth. They have tame sounding names like 'bunker-busters' but are powerful enough to break through the 300 metres of granite of the Russian Kosvinsky Kamen strategic bunker in the Urals. Chuck, it's like the Cheyenne Mountain Complex near Peterson Air Force Base, where the North American Aerospace Defense Command (NORAD) and United States Northern Command (USNORTHCOM) headquarters are now located."

Chuck looked surprised, "How do you know this stuff, a lot of it is Restricted."

"Was Restricted. At least until the Dark Web came along. Then it became part of the information freely exchanged on Google. You can literally download maps of the bunker interior now. Like I bet you can't guess where those B-61 missiles are stored in Europe?"

"What? We've got them over here?" I was genuinely surprised.

"Yes, in Belgium, Germany, Italy, the Netherlands and Turkey. About 150 of the buggers," answered Kyle, supping his beer. "Praise the Greenham Women who stood against nuclear deterrents in Berkshire. The old USAF base is common land again now."

I spoke, "Okay, but this makes it even more implausible that Simon and another pilot could steal an F-16 capable of carrying a 340-kiloton nuclear bomb from Luke Air Force Base and then crashed the plane in the desert? - It's gotta be a set-up."

The others were still silent. Then Christina spoke.

"This is a lesson in stealth," she said, "showing the powers-that-be that there is no safe place that cannot be penetrated. And that a real weapon of mass destruction could be turned in upon itself."

"Well, it keeps me in a job," said Kyle, "Worrying about all of this."

"And knowing how the other side can walk through walls," added Christina.

Okay, I said, "I've booked a table about five minutes' walk from here. It's along the Quai and around the corner and does amazing Swiss-French food."

"What are we waiting for?" said Kyle, "I'm glad to see you are keeping up your London level of action! - Bars - Food - Parties plus a twist of curious excitement."

I paid the waiter and we all trooped from the cocktail bar. Then to the restaurant, along the Quai, admiring le Jet d'Eau - the huge water fountain and ensuring all of the visitors got some classic Geneva scenes into their heads.

Then into the restaurant, which had a friendly feel rather than fancy and I could see would be a hit.

The walls were covered with pictures of typical French and Swiss scenes and there was a cupboard with many small Swiss artefacts in it. We were guided to a table which had a wooden partition behind it and a window to one side. It felt like we were secluded but in the middle of a bustling restaurant with several waiters all dressed in black outfits with white shirts.

Chantel, our waitress, asked for our drinks and pointed to a chalkboard with recommendations for the day: 'La Chasse,' it said - The hunt.

Christina kindly translated for everyone from the chalkboard, and the main menus were helpfully also in English, so we had soon ordered.

Chuck commented, "My kind of place. In Texas sometimes they bring the steaks to your table to select, but I kinda like your European varieties. Somehow the steak tastes different and some of those sauces you add. Mmm."

"Maybe it's corn-fed vs grass fed?" suggested Clare, "The cattle here roam free on the mountainside."

"Yes, but there must be something else?" said Chuck, "Or maybe I'm turning into a European!"

We all laughed.

"Have some more Faugères!" said Christina, pouring some into Chuck's glass.

"Merci!" answered Chuck.

"See, it's working!" said Christina.

More laughter, then Chuck asked, "Okay Matt, so tell us about your Swiss friends?"

I wasn't sure where to start. Maybe with those in the

apartment and then those in the Lab.

I went through, describing each person, and how I'd met them. The, one name caused everyone to stop me:

"So, this Bérénice Charbonnier, you reckon she was a girlfriend of Simon, then?" asked Clare.

"But did you ever see them together?" asked Christina.

"Whoa - slow down! Not particularly, only at the party, but I would not have guessed anything from that. Bérénice seemed to flirt with everyone, including me."

"Could Bérénice be in cahoots with Simon?" asked Chuck, "I mean the boyfriend and girlfriend thing would act as a great from of cover?"

"I'm not sure, I answered, I've also been told that Bérénice maybe plays around with other men."

"Ooh!" said Clare, "Was it a woman that told you that?"

"It was, and I'm aware how these rumours get started too.

"Okay, we need to mark down Bérénice as someone to treat with care," said Christina, "In case she is a shadow operative of Simon."

"We also need to find out who supplied Simon to Amanda," said Clare, "I mean if there was a double-agent inside MI5, then things would have become pretty sloppy."

Christina smiled, "I might be able to assist with that line of enquiry," she said.

I carried on describing people and then moved to the Lab. Amy, Rolf and Schmiddi all passed without comment, but then I got to Juliette.

"I think it's the way you describe her," said Clare, "You seem to have a finer eye when you talk about her."

"And a more descriptive turn of phrase, " added Christina.

"Yep. The way Matt's describing her, I'd say she was just his type," said Kyle, smiling.

I felt quite cornered with these responses, but then Chuck dug me out.

"Or he's already noticed she is more likely to be implicated. Remember Juliette was Levi's girlfriend and works in a sensitive part of the lab. She's also a native of Geneva and must know her ways round and probably has many contacts here. Another one we should watch out for at the party."

Then he added, "You won't get jealous, will you, Matt?"

Everyone else laughed. I thought the whole of this group seems more relaxed than when I'd seen them before, or on the conference calls. Then I realised that Amanda wasn't part of the group. I decided I would call her that evening and update her on our discussions and plans.

Chuck tinged his glass. We all looked around.

"Can I just say," he announced, "That meal was spectacular!"

There were clinked glasses and cries of 'Here, here' and '*je suis d'accord*' from around the table.

"*Plus du vin rouge,*" said Clare.

Kyle digs security

Friday Morning and I managed to get Kyle on-site at Brant. We'd waited for Amy to arrive and then asked her to help us get Kyle processed. Amy could pull rank and get Kyle through the system.

Fortunately, Kyle had his UK driving licence and a GB Passport with him. Brant held his passport in their security system when they let him in.

"I knew one day some of my own security measures would bite me in the bum, " he said as we walked in.

"I assume you've got two British passports, anyway?" I asked Kyle. He nodded, "I was hardly going to admit that though," he said.

We walked through, with Amy, to the Lab area and I introduced Kyle to Rolf and Hermann.

"Ah - so the Englishman is bringing his friends now so that he doesn't feel outnumbered?" said Hermann with a twinkle in his eye."

Kyle shook hands with both Hermann and Rolf and said, "I heard you were having trouble working out what some of the Createl code was doing? Maybe I can help?"

Hermann brought up the code on a screen. "It seems to be in this area," he said, "But if we isolate the code, then it stops working. Leave it in and everything seems to run slowly."

"I see," said Kyle, looking intently at the coding in front of him.

"It's a form of JavaScript Object Notation Web Token - A JWT - pretty standard stuff, but I can see that the token form is highly customised. It's using a token audience claim and depends upon the initial token request. It verifies that the application has been granted the permissions required to access your interface to other systems. You will need to check the scope claim in the decoded JWT's payload and make sure the permissions match."

I'd forgotten just how geeky Kyle could be. Hermann seemed to understand what Kyle was saying though. Amy looked at me and I could see her eyebrows were raised to their maximum setting.

Hermann said, "So that would account for the slow running, too. If every time Createl wants to do anything it must go through this series of checks, no wonder it runs slowly. I thought it was a protective security layer."

Kyle said, "It is really. Createl is looking for a token. If the right one is supplied, then it will run without all the tests. It knows it has the right 'audience', But if anyone else tries to run it, then it will perform sluggishly and not be capable of speeding-up. It's clever really. Levi has slugged the application to run slowly unless it is in his capable hands."

Hermann groaned, "I thought so. The result is a non-performing piece of software. No wonder the links to Selexor and to the Cyclone are so slow."

Kyle changed tack, "Had you heard the apocryphal stories about the software engineer who wrote the payroll system?"

Hermann said, "I'm not sure, go on,"

"Well, this guy named Bill wrote a payroll system for a company which worked brilliantly. Everyone was paid on time and the right tax was deducted. The company was pleased but decided they could now use a less expensive person, John, to make any further changes. Bill was let go from the company and John took over. Sure enough, the payroll worked fine at the end of the month and the company was pleased about its decision to replace expensive Bill with economical John.

"Next month came around and the system went haywire. It sent details of all the Managers' pay amounts to everyone in the company. It got the tax wrong and overpaid by 100%. Everyone's pay was late and had the wrong amounts, usually much larger."

"Wow - that was quite a mess," said Rolf, "but I'll guess

Bill had done it."

"Yes, and it was so simple. He just made the programme look for his own name in the payroll. If Bill's name wasn't there, then the payroll misbehaved."

"I see, said Hermann, "So do you think this is what Levi has done?"

"Probably, I reckon there's an identity token associated with Levi buried somewhere in the code," answered Kyle, "We find that and everything will run much faster."

Rolf said, "Kyle, thank you! You are as good as, no better even, than Matt said."

"Well, let me add my two cents, " I said, "Levi was into Image Recognition, what about if the token is some form of token associated with Levi?" I said, "This may sound crazy, but if the software recognises who is running it?"

Rolf and Hermann looked at one another.

"Amy, " said Hermann, "These two crazy Brits have worked out Levi's secret!"

"But how will we get Levi's image?" asked Rolf.

"We should put a Hi-res screen display of Levi in front of the cameras on the operating console. See if we can trick it," answered Herman.

"Better than that," answered Kyle, "If we intercept the data flow, we can probably find the token as well. Then we'll have the key to running Createl fast."

Amy smiled. Progress at last, although I could see she was looking concerned about what was about to be unleashed.

Porsche 911 4S - 132 litres luggage

Friday evening and true to her word, Juliette was waiting for me at the exit.

"So are we going over to Intermarche to pick up the drink?" she asked.

"Oh, please. Thank you so much."

I was a little concerned at the Porsche's ability to carry enough supplies, but the front space where the engine would go on most cars was surprisingly deep and we fitted plenty of bottles in there, which left the two rear seats for some boxes of beer and a few groceries. This was decidedly the bulk of what we'd need, especially if people brought something along with them.

Juliette stopped me from buying the cheese though. I realised it was more about aroma prevention than anything to do with the varieties on offer.

Then we went back to the Apartment. Juliette was familiar with it now, and suggested places to stow the party goods until the next day. She also directed me to a good Fromagerie in town from where I could pick up the cheese and a few other accessories.

There was a knock on the door mid-evening and Monsieur Stämpfli, from the Le Grande Boulangerie arrived with a couple of large boxes containing cakes and bread items from his bakery. I'd also seen a note pushed under the door from Oscar Weissbrodt. His note confirmed that the small Swiss band would be available to play throughout the afternoon and into the evening.

I was beginning to wonder whether I should have ordered some livestock as well.

Juliette asked me if my other British friends were in town yet, which I confirmed. She said she'd love to meet them, so we took off in her car for the hotel.

As I'd expected, they were sat around in the bar. I'd texted ahead to Christina and she'd rallied the troops. Not that they needed much persuading - and I think Christina must have asked them to look like a band on tour. She was in a black Russian cosmonaut tee-shirt with a huge logo, yet more skinny black jeans, a black beret and some oversized red sunglasses. Clare had a grey trilby, bright red lips and a leopard print jacket over a busy looking tee-shirt. I noticed she was also wearing the scarf that Christina had worn a few days ago.

The lads were in variations of tee shirt and skinny trousers and even Chuck had a Led Zep tee quoting Madison Gardens as the venue and a pair of Ray-Bans.

"Hi everyone!" I said, "We didn't get the rock n roll memo! And this is Juliette! Christina you've already met."

"We decided to have some fun tonight. We're going back to the bar along the front, where the waiter recognised Christina. We're going to pretend that we are with her band!" explained Clare.

Juliette looked confused by this, but Clare immediately offered her hat to Juliette, to help change her appearance.

"No - give me a few minutes," said Juliette and she disappeared towards the restrooms, with Clare.

"I was getting things ready for tomorrow," I explained, "And Juliette was helping!"

Kyle looked at me, "So what's the story, dude? And what happened to Heather?"

I briefly explained, realising that Kyle was only up to date as far as Cork.

Then Juliette and Clare reappeared. Juliette had borrowed the scarf, but not Clare's hat. She'd done something to her shirt, which, by tying a knot in one side of it made it look both off-the shoulder and more rock and roll. She'd changed her hair style too, which now looked more out-of-control.

"Look at Juliette, a perfect adaptation!" said Clare and everyone whooped and clapped.

Our waiter returned.

"I knew it," he said. He stared at Christina. "Yes - Is this your band?

He looked at me. Ah yes, I've seen you in here before as well. Are you the manager?"

I realised I hadn't pulled off the rock look.

We decided to go along to La Terrasse. It would be my third time there, in two weeks.

We hailed a couple of cabs, but Juliette walked back to her car.

"Nice wheels!" called Kyle, spotting the 4S.

"You want to ride with me?" asked Juliette. I have never seen Kyle move so fast.

Then along to the bar where we could see all of Geneva, the lake, yachts and the fountain and feel like rock stars.

Saturday Party Prep

"We gave them the rumours," said Juliette, next morning, as we prepared some coffee. Juliette had kindly driven me back and stayed over in my Apartment.

It was still early, but there was a buzz from the entry phone. I checked that we were both presentable, and then picked it up. It was Bigsy.

"Hi, Matt," he said, "I thought I'd best come around early to check out if there were any irregularities," he explained.

I buzzed him in and poured a third cup of coffee.

Bigsy had a wheely-bag, which he dropped just inside the room.

"I've brought some gear - you know, to check for anything," he explained, "Like I said on the video conference."

He unzipped the bag and produced a tracing device,

which must have been the granddaddy of the equipment he had sent to me via Amazon.

"With this I can easily check for wireless and wired-in bugs, even dormant circuits " he explained.

He worked his way around the room, pausing at spots to tear off a yellow post-it note and to place it on a surface. When he'd finished, I was surrounded by half a dozen such notes.

"There's a listening device near each of the post-it notes, " he explained.

Then he walked around with a screwdriver and quickly located each of them. Behind pictures, on lampshades, one inside of the wall switch for the light.

"Em," he said, "This is slightly embarrassing, but I think Juliette, you might also be carrying a bug or something?"

Juliette looked at him. Bigsy seemed to be staring at her chest.

"Em, I think it might be in that necklace around your neck? Would you mind removing it?" he asked matter-of-factly.

Juliette smiled, and I remembered her comment about men objectifying her.

'Here we go again,' I thought.

She delicately removed the necklace, which she said Levi

had given her.

Bigsy and I looked at it. It was delicate oval dragon locket and had 'Juliette' inscribed onto it in a flowing script.

"May I?" asked Bigsy. He pressed the sides of the locket and it sprang open. Inside was a tiny circuit board. It looked quite like a nano-SIM.

"It's a Nano SD memory card" - said Bigsy, "Levi must have given it to you for safety."

"I had no idea it even opened," said Juliette, "Can we read it?"

"We can, but I'd better copy it first, said Bigsy, retrieving a laptop computer from his now lumpy- looking bag, "Just in case we accidentally erase it."

We stared at his computer as he loaded the contents of the chip.

"It's a long key for something," he muttered as he examined it.

It suddenly dawned on me. This could be the mystery token that unlocked Createl.

"Okay," Bigsy said, "We'd better put it somewhere safe. I'm going to take it away if that is alright with you Juliette. I've made a copy of the SD contents, and I'll take that too. When we can get into the Lab, we should take this along."

Juliette nodded her agreement. Bigsy picked up the small heap of tiny devices he had found around my apartment and tipped them into a small zipper bag. Then, in another zipper bag, he placed the Nano SD card. He handed the necklace back to Juliette.

"Go on," he said, "You can still wear the dragon. It's very pretty."

"Hey Bigsy," I asked, "Would you like to stay here now and help us get ready?"

"No, Look I'll be back as soon as I can, but I need to drop all of his stuff back in my hotel room, away from prying eyes. I should be back in about an hour though."

PART TWO

Another girl another planet

You get under my skin
I don't find it irritating
You always play to win
But I won't need rehabilitating
Oh no
I think I'm on another world with you
With you
I'm on another planet with you
With you
Another girl, another planet
Another girl, another planet

Space travel's in my blood
There ain't nothing I can do about it
Long journeys wear me out but
I know I can't live without it
Oh no
I think I'm on another world with you
With you
I'm on another planet with you
With you
Another girl, who's loving you now
Another planet, who's holding you down
Another planet

= Peter Albert Neil Perrett

Party

We'd just finished setting up the kitchen as a makeshift bar when the doorbell rang. It was obviously an internal visitor from another one of the apartments.

I answered the door and there stood Bradley Floyd and his partner, Jennifer Hansen. I'd not seen them since the time that Simon disappeared and wondered if they'd know more.

"Hey, its great to see you," said Bradley, who peered into the apartment. "Yours is just like ours! He exclaimed, except for the distance from the ground!"

"Happy Dèsalpe, "said Jennifer and thrust a large rectangular tray into my hands. They are cookies, home-made, the American way!" she said.

"That's perfectly lovely!" I exclaimed and I secretly wondered if they had any American style hash in them.

"No, they are plain chocolate, no student-style adaptations!" laughed Jennifer, apparently reading my mind.

"And a couple of grown-up bottles of booze, "added Bradley, handing me a bottle of Yukon Jack and another of Jack Daniels, "I suggest you hide the Yukon Jack and I'll come around to share it with you another day!" he winked.

Another tap and Bérénice Charbonnier was at the door. I hadn't seen her since the fateful evening of Simon's disappearance, and I wondered how she had adjusted to life without Simon. She certainly seemed bubbly. "Come on, Mr Bise man! She offered me both cheeks, and I did the French-style greeting that she had taught me at my first party in Geneva, I could feel her surprisingly powerful body as she manoeuvred me around.

"Oh, and this is Niklaus, Niklaus Zeiler." She didn't explain further, but I assumed he was her latest conquest. Niklaus was a tall and wiry man, early thirties, piercing blue eyes, short brown hair, and a short, stubble-like beard. He was wearing a blue close-fitting sweater and wore a red string necklace which looked as if it could have had dog-tags on it.

"Hi," said Niklaus, "I've been hearing about you - the scientist in the block! But a very sociable scientist, by all accounts." He was carrying a box of beer and placed it gently on the table. It seemed too early to ask what he did, but Bérénice was quick to explain. "Niklaus works on the boats, from across the river. If you ever want something special, you should talk to Niklaus."

Niklaus smiled, "Yes I do that in the summertime. In the winter I go up onto the slopes, to teach ski-ing!"

"Oh yes," I asked, "and which area?"

Niklaus answered, "From Geneva there are so many choices all about an hour away. I am usually based in one of the villages of Les Portes du Soleil. There are nine different resorts and over 650 km of pistes and is easy for me and for tourists to get to from Geneva. It also has some of the most dependable snow. It's partly in Switzerland and partly in France too, has plenty of families and non-skiers, so there is always someone to teach at any level. There's over 1,000 instructors there, so you know we are on to something!"

"Well, we are celebrating the cattle coming off the mountains today, so it can't be that long until the season starts again!"

Niklaus answered, "As a matter of fact, the snow usually starts in a small way in September already and then gradually builds through to November. I usually pack my bags in November through until around the beginning of April."

"Are you going to go along?" I asked Bérénice, "I guess everyone in Switzerland can ski?"

"Oh yes, I'll be there enjoying the skiing, when the time comes," answered Bérénice.

A buzz at the main door and it looked as if the gang from the Four Seasons had arrived. They seemed to be carrying on with the rock-chic look.

"Hello Matt!" boomed Chuck as he entered the room,

immediately recognising the tones of another American and making his way over to introduce himself to Bradley.

Clare and Christina had crossed the room with Christina talking to Juliette and Clare to Bérénice. I could notice Christina looking at Bérénice as if weighing her up.

Bigsy was looking at the kitchen table with its supplies of drink and chatting lightly to Kyle.

Then the front door again and this time it was Aude, "Wow, you have certainly got to know some people!" she said, amused to see so many people present, at which point the door rang again and Oscar arrived with four of the band members. They had brought traditional instruments but also seemed to have a couple of amplifiers and electric guitars.

Oscar smiled, "I think you'll get your share of Swiss traditional music now!"

Bigsy looked delighted because he could now help the band set up their equipment and assist with the sound check.

The Apartment was getting full by this point and quite noisy, but I realised I'd invited all the neighbours around, so I thought I was in the clear. Then the door again and Hermann and Rolf arrived, clutching bottles of Schnapps.

"Er - we invited Amy as well," explained Rolf, as Amy appeared from behind them, clutching a plastic tub containing an Apfelstreudel.

"I hope it's okay!" she said sheepishly.

"The more the merrier, I said, "Although what goes on in Rue de la Confédération stays in Rue de la Confédération!" I said.

Chuck tinged a glass and said, "Speech Matt!"

I couldn't think of anything, so I said "Eat, Drink and be Merry! Mange bois et sois heureux!"

Everyone laughed and I thought I was off the hook.

"Your French!" said Juliette, "It's improving."

"I know, although I think heureux means happy," I said.

I looked over at Oskar, who introduced me to the band. "This is ever so kind of you all!" I said,

"It's so good to have a visitor who wants to make an effort!" came the reply.

Oscar said, "Yes, and otherwise they'd all be in the bar this afternoon, anyway!"

The band started up with Ranz des Vaches - which they told me was the Gathering of the Cows. Then Der Chüjerstamme, which included some proper yodelling. A couple of songs later I realised that most of the songs being sung were in German and I asked why.

"There's more Swiss folklore in German than French, but

it doesn't really matter, we'll sing in whatever language is needed!"

"Look - we've another musician in the house!" called Chuck, "Can we get a song from Christina?"

Christina looked sheepish and then over to the band leader. He grinned and said, "Sure, we can do a pop number as well, if you like." He pointed toward their two electric guitars.

Christina nodded, "Okay, Let's try something. Do you know anything by The Rolling Stones?"

The mandolin player sketched out the riff on his mandolin and then moved across to the electric guitar. Jumping Jack Flash appeared and Christina picked up the microphone and sang it with due force.

"Wow," said Kyle, "I loved it!"

"Can I borrow the Telecaster for a moment?" asked Christina.

She plucked out the first few notes of Satisfaction, another Stones' tune, which she sang as a slow guitar and voice number, but each time when she got to the chorus, she left it out. "Cat Power..," she trailed at the end.

The Swiss band applauded, and the band leader said, "How about Nirvana?"

I had to adjust my reality controls at this point. A Swiss oom-pah band asking Christina Nott to play Nirvana, was completely uncharted territory.

"You know I used Teen Spirit on my last tour?" asked Christina, "For when I was in the United States and especially in Seattle?"

"I read about it," said the band leader. "But I never thought I'd get to hear it."

"Guys, it's Em-A-G-C-Em" he said, reducing one of Nirvana's greatest hits to a 5-note chord sequence.

The euphonium player picked up the bass notes Em-A-G-C, then Em-A-G-C again. Then Christina started singing and picking at the Telecaster - she was treating it as a performance and doing some of the moves.

Load up on guns
  Bring your friends
  It's fun to lose
  And to pretend
  She's o-ver-bored
  Self-assured
  Oh no, I know a dirty word

Clare joined Christina at the microphone

Hello, hello, hello, how low
Hello, hello, hello, how low
Hello, hello, hello, how low
Hello, hello, hello

Then, to everyone's amusement - the band leader cut in for the next verse, which he duetted with Christina.

With the lights out
It's less dangerous
Here we are now
Entertain us
I feel stupid
And contagious
Here we are now
Entertain us
A mulatto
An albino
A mosquito
My libido
Yeah

Christina flipped the controls on the guitar and went loud for a piece. The euphonium answered back. This was a strange performance but kind of smokin!

The rest of the room was wrapt in the song. Amy looked as if she was about to explode with delight.

Clare had looked around and picked up two table napkins which she was using as cheerleader pom poms, like they did on the video.

Two more verses and choruses and the band brought the song into rest. Christina whooped a couple of yodels at the end.

Everyone clapped everyone else. It had been a moment. I think Christina with the oom-pah band had just made my party. Christina was profusely thanking the band for letting her play with them. The guitarist was trying to get her to sign the Telecaster with a Sharpie.

Amy walked over to me. "How do you know her? She's fantastic!" I called Christina over.

"Amy really enjoyed that," I said, "I think she has questions!"

The party was ignited now, I could just let it burn through.

I Cant Get No Satisfaction

Capo 5 **Em G Am C [Play slow]**

When Im drivin in my car
And the man come on the radio
Hes tellin me more and more
About some useless information
Tryin mess my imagination

When Im watchin my tv
And a man comes on to tell me
How white my shirts can be
But, he cant be a man cause he doesnt smoke
The same cigarettes as me

When Im ridin around the globe
And Im doin this and Im signin that
And Im tryin to make some boy
Baby, baby, baby, come back
cause you see
Im on a losing streak

When Im ridin a round the globe
And Im doin this and Im signin that
And Im tryin
And Im tryin

Adaptation - Cat Power
Lyrics Mick Jagger / Keith Richards

Sunday satisfaction

Juliette and I were still at mine until Sunday morning.

The party had finished at around 2 am, which I felt was respectably late. The band had left around 10 pm and then the local guests had mainly gone by midnight, so it was just the out-of-towners left behind. They eventually got the hint and slid away to walk back across the bridge to their hotel.

I'd arranged to meet them there around lunch time for a debrief before they headed back to London. Chuck told me their checkout time was 11am, although the flight wasn't until the evening.

We decided to meet in the cocktail bar (again). This was pure practicality, so they could check their bags into the left luggage at the hotel and then be picked up later in the day to go back to Geneva airport.

Juliette said she'd skip the session in the hotel and make her way home, to get her head ready for Monday. I didn't say anything, but I thought this was useful so that any review or thoughts about people in the Lab could be discussed quite openly.

Juliette left, and I finished clearing up the mess. I'd gone through four big bin bags full of rubbish and could now see all the flat surfaces in the apartment again. There did seem to be a huge number of wine bottles around, far more than I'd remembered buying, but then I realised that the local folk could simply pop back to their own apartments to replenish stocks.

Most of the food had gone, but I'd noticed the band making serious inroads to the Strudel and the chocolate cookies and brownies. No wonder everyone was buzzing; there must have been a serious sugar overload.

Once I'd taken the black bags to the big bins in the yard downstairs, I decided I'd walk over to the hotel to blow the cobwebs out of my head. I wasn't hung over like that time with Rolf and Hermann, but I knew I was super-tired. I really needed to get back into bed, but I decided that's what I'd do once I'd said goodbye to everyone from England.

Inside the bar, the waiter greeted me, I knew this was maybe the fourth time I'd been here and each time it had been with a group of others. This time I was first and I sat down at 'the usual' table and ordered a coffee.

Clare was next to arrive. She'd checked out and parked her luggage with the concierge. She looked fresh as a daisy. "Great party last night!" she said, "What a brilliant way to gather information."

Then, Chuck and Kyle arrived together. Kyle had spotted Chuck in the checkout line, and although Kyle wasn't checking out until midweek, he'd kept Chuck company.

"What- drinking coffee?" gasped Kyle, as he ordered a beer. Chuck did the same.

"Great party!" said Chuck, "and nice to meet those two from Charlotte - Bradley and Jennifer. They both seemed to be lovely people."

Then Christina arrived, with Amy. I had, once more, to adjust my reality goggles.

"Hey Amy," I said, "I wasn't expecting to see you today!"

"No, I can't stop, I need to get back to my own apartment," said Amy, "But Christina persuaded me to pop my head into the bar to see you were all functional this morning!"

Amy was just about to leave the bar, when Bigsy walked in. He looked like I had done after my first meeting with Rolf and Hermann. "Hi Bigsy, are you okay?" asked Clare, looking worried.

"Hermann and Rolf," he gasped, "They can pack it away!" He saw my coffee, "I think I need one of those and -er - something bread-like to eat."

Clare fished around in her bag, "Here," she said, "Two paracetamol and you can have my water - I suggest you drink it all."

Bigsy meekly complied, although even his movements to drink the water seems laboured.

"So what did we get?" asked Clare.

"I wasn't entirely convinced by Niklaus Zeiler," said Christina, "All that stuff about the ski-slopes outside Geneva came across a little bit too much like a guidebook."

"Yes, but he had that sort of fit, toned body, that one associates with a ski-instructor," observed Clare.

"But look around Geneva," said Chuck, "Half the men here look as if they go swimming or skiing every weekend,"

"I had noticed," smiled Clare, looking over toward the two skinny waiters chatting at the bar.

"I tried him out on the languages too," said Christina, but I mixed a little bit of Russian in to see if he would notice. It didn't seem to phase him."

"Where does Niklaus live?" I asked.

"He's just moved into the apartments. Not Simon's old one, but one upstairs on Bérénice's floor.

"Wow - she doesn't waste much time!" said Bigsy; the paracetamol was obviously having its effect.

"Maybe," said Clare, unless it was all a convenient set-up?"

"That brings us on to Bérénice Charbonnier," said Chuck, "What did anyone gather?"

"She was surprisingly difficult to talk to," said Bigsy, "Like she was already suspicious of me,"

"It must be your winning ways!" grinned Clare.

"I asked her about the Geneva area as a way to lead into her background," said Bigsy, "But she shut down when taking about brothers and sisters, or where she was from.

"Actually, I thought she had a really realistic British accent. If I'd met her in London, until I knew her name, I'm not even sure I'd have noticed she was 'a European'."

"I thought you Brits were still European," laughed Chuck, "Oh no, I was forgetting the Channel"

"I think I got some more from her," said Christina, "I was talking about swimming and mentioned that I did scuba when she got quite interested. I mentioned the Red Sea, which is one of the places I'd been diving and Bérénice knew all about it."

"She was talking about Sha'ab Abu Nuhas which is a submerged reef close to the busy shipping route to the Suez Canal. She had dived to the Carnatic, which is the oldest of the wrecks there - a cargo and passenger steamer split in half that looks almost like the coral reef around it now. I made a mental note that she must be quite and experienced diver to have been out to that in among the stingrays and giant moray eels."

"So, are there any scuba dive spots on Lake Geneva?" I asked.

"All around it, apparently," said Christina, " Even wrecks to see!"

"Is a picture building up in anyone's mind?" I asked.

"Then Amy? " said Chuck,

Christina spoke again, "Ah, Amy van der Leiden, from Amsterdam. We talked about Amsterdam and it was obvious she'd lived there for a long time. I was there for a couple of years when I was making music. I get the impression that Amy is a bit of a fish out of water here in Geneva. She's holding it together but doesn't like the direction the research is headed. Amy seems to think that her boss and the senior management are just in it to take profits. She seems to be in over her head though. It seemed like a simple enough story, and basically true and heartfelt."

"I was with Rolf and Hermann for quite a piece of the evening," said Bigsy,

"And so was I," said Kyle, "I guess we talked far too much about Information Technology and security."

"Rolf is the real deal, though," said Kyle, "He knows his security systems and a lot about system development - I'd say he was an earnest researcher."

"Did you think the same for Hermann?" asked Bigsy,

"Only I thought he wasn't quite as sure of the detail compared with Rolf."

"Yes, I agree, but the two of them together are a great team - a Yin and Yang with complementary strengths. Hermann will have sometimes quite innovative ideas, but then Rolf knows how to make them happen."

"Do you think they were on the level?" asked Chuck,

"Oh yes, they both seem to be very straightforward people. Well, aside from being hard drinkers, that is," replied Bigsy.

"I spent a decent amount of time with Bradley Floyd and his partner Jennifer Hansen," said Chuck, "They seem pleased to have another American to rib, so we got on well. One thing I noticed was their encyclopaedic knowledge of Charlotte. I'd say their story holds good."

"It was the banking that intrigued me...Say Matt, weren't you in on that discussion as well?"

I was and remembered that Bradley was no ordinary banker. His company seemed to be day traders and he said he was looking for extra profits from Brant. I remembered how he had explained it:

"Well, last week, Brant's shares gapped down more than 5% on the open, relative to the prior closing price, and continued to fall. It was because of some leaked overnight news about the Artificial Intelligence Systems bought from Createl running slowly."

Chuck asked, "So where could someone like Bradley get that news from?"

"That's what I wondered, because there were only five, or maybe six of us that knew about the software problem," I said.

Bradley had continued, "Soon it was down 8%, but then it started to rally. Some investors who were thinking about buying the stock jumped in, thinking the stock was a bargain at a 5% to 8% discount."

"That's an artificial rise caused by less well-informed people buying in at what they think is a discounted price," I said.

Then Bradley had continued, "Then the day traders who went short early in the morning decide to cover their short positions, which helped to fuel the bounce. Soon, the price was back near where it opened, but remember it is still down 5% on the previous day. Now the price plummets lower once again, as those who hadn't sold their shares that morning are relieved to do so near the opening price - to get out while they can."

Bradley had explained, "In the trading business we call it a dead cat bounce. We usually try to make money on it by shorting it. - In other words, selling shares in the expectation we can buy them back at a lower price later in the day."

Chuck said, "I worked out that a dead cat bounce must be when the price gaps down 5% or more, continues to decline after the open, but then has a false rally."

Chuck added, "and why short? Because the 'cat is still dead.' Just because the stock bounced doesn't mean it's going to keep surging. Significant damage was done to the stock price on the bad software news, so the underlying issues are still there, and real investors are scared. A bounce is a second chance for those scared investors to unload shares at near to the opening price, but all the time, pushing the share price lower."

"Let's just think about this for a moment. Bradley had a fun party, invites along everyone from the apartment block, which includes several Brant employees. Then he gleans the information about Brant's problems and uses it as an insider trading mechanism to tweak the price of the shares."

"Okay," said Clare, "I can see the logic of this, but I can't see that this would also involve murderous crimes?"

"Agreed," said Chuck, "It just illustrates that plain-living Bradley might be a ruthless day-trader."

"But what about his wife?" asked Bigsy.

"You mean Jennifer Hansen, his partner?" asked Clare, "I didn't get the feeling that she was working? Did anyone else?"

"No, not really, although I did find out that she was a good friend of Bérénice," said Clare.

"Hmm. That's Bérénice, the reporter for Le genevois - The Genevan. Quite handy to filter rumours into a mainstream media source, " said Christina.

"Okay, said Clare, "So we have Bradley and Jennifer running a party, finding insider information about Brant, providing it Bérénice at The Genevan to publish whilst all the while trading in Brant's shares."

"And we have Simon's erstwhile girlfriend Bérénice, a fully trained scuba diver, and Niklaus, with access to boats, operating around the lake when Levi's boat mysteriously becomes adrift, and Levi falls overboard."

Christina looked around the group, "Run that another way and you have a horror story where Simon is on the boat with Levi. Simon overpowers Levi, pushes Levi into the water, where Bérénice is already hidden in her scuba gear. She drags Levi down. Then Simon and Bérénice transfer to Niklaus's craft leaving Levi in the water and the yacht becalmed." said Christina, "Brutal, yet mysterious,"

Everyone looked shocked at Christina's take on the events.

"I had a good chat with Oscar, too," said Kyle, "I didn't realise he was a priest until we'd been talking for about 20 minutes. Even then I had to ask him a direct question! We were talking about music most of the time and he seemed to like plenty of the same British bands as me. I asked him about life in Geneva and that's when he started to talk about his parishioners and so forth. He seems to be on the level, in my humble opinion."

"I agree," said Clare, "And he seemed to care a lot for the

people in the apartments and said that Bradley and Jennifer were his particular friends, both earnest evangelical folk back in North Carolina."

Chuck nodded, "Oh yes, I should have mentioned that. Clearly, they were heavily influenced by their time in 'The Bible Belt'. North Carolina is in the top ten most religious states, and some even say that Charlotte is the Capital of the Bible Belt - although I, like many others, think it is Oklahoma City - but then I'm from Texas and Oklahoma is much closer!"

"What about Juliette?" asked Clare, 'Matt - you must have an opinion there?"

I decided I couldn't hide our relationship.

"Juliette and I are quite close. She's a psychologist by trade and was the girlfriend of Levi, so in that way is quite connected. To begin with, she was quite guarded with me, and I think that's how she will come across to anyone else as well. She's originally from Geneva but has travelled and knows her stuff regarding HCI (Human Computer Interfaces). I would guess that Brant have her marked out as a high-flyer."

"I also noticed Juliette's evasiveness; I wondered if it was about me, " said Kyle.

"She is the same with women," said Clare, "I think it would take a while to get to know Juliette,"

"Juliette seems to have a digital gift from Levi too," said Bigsy, "We discovered it last Friday, I really want to try it in the Lab. To see what is in the file."

Kyle looked at Bigsy, "I'm here until mid-week," he said, "If you could stay on an extra couple of days, we could check out whatever Juliette has found together?"

Bigsy looked at Kyle, "Are you sure - that you can stay on, Kyle? I'll change my room booking in that case."

"We haven't talked much about Aude," said Christina, "And I think she only stayed for about half an hour."

"I get it, though, if she's the representative for many apartment blocks then she could probably spend all her time at parties. I guess she showed up and said 'Hello,' to everyone like a true Real Estate pro." said Clare.

"And we didn't mention the band! They all seemed like regular upstanding Genevois citizens, musicians and churchgoers - I enjoyed chatting to them actually," said Christina, "and we talked about all kinds of music, way beyond the kind that had cow-bells in it."

Kyle said, "That other tune you played on the mandolin! - Brilliant!"

Christina replied, "Thanks but the band loaned me the mandolin, which I could play like a balalaika! Normally they are tuned like violins, so I had to retune it and pretend it only had 6 strings instead of 8! I used to know plenty of Russian folk tunes, so that was what you got. Kalinka, Katjuschka, and Krassnyi Sarafan (The Red Sarafan!) I didn't want to take over the party!"

"But you had to play so fast!" said Kyle.

"I know, when I gave up the balalaika it was for the guitar, where everything seemed so much slower."

Clare interrupted," Now we've got some good intel about some of the guests, heavily laced with supposition. Here, in order of notoriety:

- Simon - we think was a super spy and an assassin, but don't know who he worked for.
- Bérénice - his accomplice and an accomplished scuba diver.
- Niklaus - a possible Simon replacement who was also implicated in watery Levi's death.
- Bradley and Jennifer - clandestinely insider dealing in Brant's shares.

The others all seem to be what they claim to be.

- Amy in charge but in it over her head, maybe threatened by Kjeld Nikolajsen.
- Hermann - Bright, if somewhat boozy analyst
- Rolf - strong technician and a great buddy of Hermann
- Juliette - Difficult to get to know (unless you are Matt)
- Aude - A social butterfly.
- Oscar - An unlikely, but well-liked priest

"I wonder if they expected us to be listing them all out on a sheet of paper after the party?" smiled Bigsy, "Let alone that they would get some of Clare's patented 'one-liners!'"

Clare took a photo of the sheet of A4 paper onto her phone.

"I'll send it to each of you - but PLEASE don't just leave it in your in-trays!" said Clare, "Best send it to Amanda, Jake and Grace too," said Bigsy, "Only add their last names to that version to give Grace a chance to dig deeper."

Monday at the Lab

I'd arranged to meet Kyle at the bus stop, and he arrived with Bigsy.

"This takes me back to waiting for buses around Gloucester Road," said Kyle.

"I know that area too," said Bigsy, "We used to go to pubs around there, let's see now...The Drayton, oh and that one on Onslow Gardens - The Anglesea Arms? And that other one near the station. Er.... The Stanhope? Always crowded."

"Yes - I think all of those that you listed are pretty busy," said Kyle, "I reckon if I was looking for somewhere quieter, of those I'd go to the Drayton Arms."

"Yes, it's just enough off the track from the tube, isn't it!"

I was wondering if the day would become more surreal than standing waiting for a company bus in Geneva, talking about Kensington boozers.

Our bus arrived on time, and we took the now familiar route to the campus. I'd worked out that it might be difficult to get Bigsy into the Lab, so once again I'd prevailed upon Amy to assist. I'd also reminded Bigsy to bring his passport and driving licence as forms of identity.

Kyle, on the other hand, was now badged until midweek.

Fortunately, Amy was already in the Lab and came through to book Bigsy in. She welcomed him and said she had thought he was travelling back on Sunday.

"That was before we discovered a new key which may help move your project along," explained Bigsy. I noticed he didn't mention where the 'key' had come from.

Key

I took Bigsy and Kyle through to the Lab, where Rolf and Hermann were sitting discussing another software idea.

"Hello Guys," said Hermann, "Now I think you Brits are trying to out-number us!"

Juliette appeared from another experiment pod and Kyle said, "We think we have something that could speed up the processing of the Createl system,"

Herman and Rolf both looked very interested. "It's a long sequence which we have derived and which we think Levi would have used to bypass all of those additional checks during processing."

Rolf asked, "But how do we feed this into the Createl system?"

Bigsy looked at the hardware being used, "It's an Exascale, isn't it? These are still hard to come by."

"Not here," said Hermann, "We seem to be able to use

them for everything compute intensive."

Bigsy looked at the console. "Hah, but it is reassuringly low-tech to bootstrap these devices," he said, "They still use a thin hardware interface and, guess what,, it inherits many of the original DOS and even CP/M conventions."

Bigsy tapped some commands into the console and reduced the system to running a command-line interface.

"See," he said, "we are down at the metal now. He typed in a net command and the system started listing drive letters.

"Incredible, we have one of the fastest computers on the planet here, but it still looks for a floppy disk drive - A:"

Rolf smiled, "Oh yes, and many of the old DOS commands still work when you are using the system in such a basic mode."

Bigsy typed in "Eject d:" and a small tray slid out from the front of the console.

"Well, that's interesting, " said Bigsy.

"I've never seen that before," said Hermann.

"Me neither," said Rolf.

"It looks as if this hardware comes configured to read memory cards, presumably to feed in some extra parameters," said Bigsy.

Bigsy added, "My guess is that the original configured Nano-SD card is just such a set of parameters."

Kyle was also looking at the console. "I think you could be right," he said, "Bigsy - have you made a copy of this SD card?"

"Yes, but it was to a computer file, not how it is formatted on this SD card," answered Bigsy.

"Okay, just a minute," Kyle looked inside his bag and produced a small black cube, "Its a memory chip copier, nothing fancy, but it will make an exact copy of whatever is on that little card."

Bigsy looked at the devices, "Pretty simple, where did you get it?" he asked.

"Oh, from Amazon," answered Kyle. He plugged it in to a spare USB socket and the unit flicked to life, "Now we need a spare SD card."

Rolf walked across that Lab, " That we have...hundreds of them," he produced a cardboard box which had a selection of brand-new plastic encapsulated memory chips. Bigsy looked through it until he found one suitable.

"This one, it's the same as the one we want to copy."

Bigsy and Kyle did some pressing of tiny memory chips into Kyle's black box and then a few seconds later Kyle announced that the original chip had been copied.

"Now we'll put it into the console!"

Bigsy placed it into the small tray that had ejected and then typed in "Load d:" and everyone watched the tiny card being sucked into the computer.

Then Bigsy typed another command "Run" and the Exascale computer began its normal sequence. Several second later it was at a familiar screen.

"Oh Wow!" said Rolf, "If that is really the start-up screen, then we have reached it in seconds instead of minutes. I think the 'key' is working and we are running many times faster than normal.

Rolf ran the test harness to check that the Createl system was still functioning.

"Oh my God," he said, "That was so fast. That test was over in seconds but usually takes over 40 minutes."

Hermann asked, "Can we try the big Test Suite?"

Rolf nodded agreement and I saw him tapping a few more command into the console.

Hermann looked at Rolf's face, "What is the matter, won't it work?" he asked.

"No, it has run and done everything! The took a few seconds instead of three hours!"

"We have to try the Cyclone again then," said Hermann, "Are you prepared to do it?"

"Of course, " I answered, wondering whether either I or the rat would notice the performance hike.

Rat Vision

Rolf and Hermann meticulously prepared the Cyclone and we placed it on my head. I could see that Kyle was fascinated by the whole process, although he did stifle laughter when he saw the other end of the wiring appeared to terminate with the rat's backpack wiring.

Now we had rebuilt the chain that I'd tried a few days ago. Matt to Createl to Selexor to Rat, all running in an Exascale computer, but now with an added key from Levi.

Rolf completed the systems checks, asserted that there was no direct connection between me and the rig, then he started a couple of video cameras and asked me if I was ready.

"Yes, let's go!" I uttered, and then once again I had the sensation of falling from a high place, this time so sudden and fast that I had to grab the arms of the chair in which I was seated. I could remember the original event being laboured and slow, but this time I felt oiled and fresh. I knew I was inhabiting the white Rat, and could even feel the heart rate rising, but this time it was of a separate autonomous being, rather than fatefully linked to my own biomechanics.

We walked around the cage, and I found the food supply, which Hermann had provided. It seemed to comprise chocolate buttons and which my rat-side ate with great relish.

All my vision was receiving an overload signal. It was like an infinity wall of lights, shifting around in any blank spaces. I realised it must be data, but I had no way to interpret it and I wondered if either my brain or the augmented Exabrain would, if left long enough, start to be able to process it, like the way human brains know sufficient to flip the images on eyeballs so that everything appears the right way up. I recollected that scientist Dr. George Stratton had experimented on himself and his own brain transition took five full days to become functional.

This was interesting, because at this new faster system rate, I could continue to think, whilst knowing of my own identity and that of the rat. It was like Levi's key had unlocked the potential of the device.

I noticed some movement above the cage, and once again a second black rat was introduced. Larger than me, it was attempting to be the alpha and to take the food. This time, instead of positioning for a fight to the death, we started a rough and tumble fight, like two cartoon characters. I worked out that my rat's autonomous behaviour was taking over and that this was what rats must do. Fight, but in a playful way, without drawing blood. The rats were behaving socially toward one another.

I did notice that the 'data dots' changed style during the play fight, and I seemed to be surrounded by thousands of tiny fireflies. The other rat retreated, carrying with it a sliver of the food.

I was going to follow, but instead found my rat was pulling the food in the opposite direction, and onward to a hoard in the corner of the cage, which I recognised also contained a stack of lab blocks of rodent food. I was a hoarder! I could not override this behaviour, which I felt must have been as instinctive as moving away from fire.

Then a collapse of the infinity wall. I could see normally again and realised that Rolf must have switched off the connection.

I looked at my body and realised I'd been sweating extravagantly from the exertion.

Kyle spoke, "Well done, mate, that was spectacular! Although I think you might need a shower!"

Juliette looked at me, "Incredible," she said, "I think we have just witnessed the first generation of a new HCI!"

Rolf and Hermann were chatting away animatedly in German, but then Rolf noticed me looking over and said, "Matt, aside from your heat output, you seemed to run at a normal breathing and heart rate this time. I had the system set to interrupt if your heart went over 176 bpm. You were only just over 100 for most of that."

"We'll have to keep Levi's Key a secret," said Kyle, "Now that we know it has such a fundamental impact on the way things work."

"Shall we tell Amy about this?" asked Juliette, looking across to Rolf and Hermann.

"Yes, I think we should, but not the Key" said Hermann, "Although she will want to know what we have changed."

Rolf smiled, "Ach Ja - the configuration settings, of course," he said, "I can print out a few pages of hexadecimal to show her. I can explain that we swapped the Afferent and Efferent settings in case we had got them mixed up - you know, the input and output."

I asked, "But surely there still a long way to go before this can be made usable? I mean, it's massive and unwieldy - running on an Exascale computer and all. Plus, look at all the cables into the headgear. And that rat has to carry a backpack as well."

"We can come clean. We are only using a tiny proportion of the Exascale," murmured Rolf, "We asked for such a brute of a computers as part of a budget grab - It's like having the best car in the executive car park."

"Boys and their toys," smiled Juliette, "Amy knows too, but she was prepared to go along with it to get some attention from Brant. They know this must be serious work if we are using Exascale computers."

"I reckon we could run the whole of Levi's system on a Windows 10 system, but by saying it needs an Exascale, it sure puts a lot of competitors off trying!" said Hermann.

"We'll let you get freshened up and then hold a debrief, " said Rolf.

"Here," said Juliette, she threw one of her tee shirts across.

"It's one of my spare tops," she explained.

Debrief

I returned to the Lab, wearing the powder blue tee shirt from Juliette. It seemed to have a small flag on it, which I thought was of Norway, and I decided that it must be a generic tourist tee, because there didn't seem to be any adaptations to the fit. It was just a little - er - tight on me.

"It's from Norway," explained Juliette., "One of my favourite places, especially the crinkly edges!"

Luckily, I'd got my jacket too, which I could wear over the top and which gave me a more-or-less integrated look.

"Hey - that colour suits you!" quipped Kyle.

Juliette looked on in amusement and Rolf and Hermann were wriggling in their chairs anxious to begin the debrief. I noticed Amy also, who we had invited to the meeting.

"We looked at the video recording of the experiment whilst you were gone," explained Amy, Mighty impressive! Rolf tells me he swapped over some of the code, and everything went faster,"

"That's right. My senses merged with those of the rat, and I think the increase in speed meant that I was able to do more. The rat also sent back its own responses and I found that there were a couple of levels - like my personality was transportable to the rat, but the rat's ultimate presence doing things was almost running at another level. I'll call them Persona and Presence."

"The Presence was running the automatic limbic system levels of operation, like fight and flight, plus the brain stem functions, like breathing, pulse and blood pressure. I couldn't adjust these much, but it was better than when I was underpowered in the prior experiment and my pulse peaked above 200 bpm. This time Hermann had installed a cut-out if I went over 176. I didn't get close."

"Interesting," said Amy, "The more autonomous functions seem harder to control. Do we think this is a lack of processing power?"

"I don't know, but the other more deterministic part of my brain, I'll call it the Persona, was capable of modification, at least until the instincts from the Presence came along and over-rode it."

"To be honest, I don't think I've really learned how to drive this thing yet, I'm sure my own brain needs time to adapt."

"Oh yes, and there were many swirls of light in my vision, away from the thing I was directly looking at. It reminded me of an infinity mirror, but perhaps the dots and swirls were some forms of uninterpreted data. It's a very trippy experience, actually."

"What about control?" asked Amy, "Did you have any?"

"I did, and I could tell that I had control as well, at least until the rat limbic reactions overtook me. I could also think about next actions. But I'd still say that the rat's inbuilt instincts could overpower my own brain's responses."

"That could also be a language thing," said Hermann, "We are using electronic stimulus which isn't the same as the mixed stimulus that a brain receives. Think of it as dialects, ours is all-electric but the real one is a blend of electrical, chemical and hormonal."

"Yes," said Rolf, "and you are also having to 'talk rat' - I guess rat-to-rat or human-to-human communication would work better."

"Okay," said Amy, "This is good, and I already have sufficient to take a good news message to Kjeld Nikolajsen. He has been on my back for days about our need to make some progress."

Amy departed, and we decided she was heading straight for the third floor to speak to Kjeld.

"Okay, we need to talk tactics for a few minutes," said Kyle, "You all know that my specialism is security and I'd liked to propose a few measures around here."

"Firstly, we need to ensure there is a barrier protection for the device. That's at least two layers. One is the external room security. You should get all the cards reprogrammed so that generic access to this area is no longer permitted."

"Second, you should add your own extra layer of security. Get a non-approved key system and have it wired in. Issue the keys only to lab users."

"Then you should put security on to the relevant computer systems. An extra layer. That seems to be what Levi had done, but you'll need to add something of your own that slows people down and makes it difficult to access the main system."

"Additionally, you should hide the real system. Put another one on display but wire the real one away into a locked closet somewhere. If anyone is going to steal anything, you want to give them a realistic looking dummy."

Rolf nodded, He'd been thinking about each step, and they were all achievable.

"Come and take a look, " he said, he guided us toward a cleaning cupboard and then pulled open its large internal cupboard.

"These labs were all built to contain supervised substances, something which is very necessary in the labs which hold drugs like cocaine, used in some of the experiments. But we don't have any chemicals beyond the rodent food," he gestured to several sacks laying in a corner, "So the cupboard here is empty. Of course, it is well provisioned with power and extractor fans, but I don't think we've ever used it. Until now when we can install one of the Exascales in it. The one with Levi's key."

Kyle nodded his approval. "Yes, that is good and with the various measures I've described it should prevent theft."

"But do we even know of anyone who would come along here to try to steal anything?" asked Juliette.

"No, but once Kjeld knows about it, I won't be surprised if word travels fast."

Monday evening.

Bigsy and Kyle were going to be in town for a couple more days, so I suggested to them they could go for a quiet beer in the MO, and I'd join them later. From where the bus dropped them off, it was practically on the way back to their hotel, in any case.

After they'd left, I mentioned it to Schmiddi and Rolf as well, so there would be a little gang present. Hermann and Rolf didn't need persuading and a few minutes later I could see the pair of the getting into Schmiddi's car to head for Central Geneva. With Schmiddi's driving I wondered if they would be there before Bigsy and Kyle.

I hitched a ride with Juliette, and we drove back to my Apartment first, so that I could change out of her tee-shirt into something of my own.

Juliette poured us both a white wine, "So who are you really?" she asked.

"I can tell you are not quite what you seem, and you seem to know too many special people too. That Christina isn't just a pop singer, I can see that she knows how to handle herself as well as speaking about a half dozen languages including Russian.

"Then there's Chuck, an American military man, if ever I saw one. Probably at least a Colonel too."

"And even Bigsy and Kyle are well past the normal range of computer specialists that we have on the base. They managed to get into that Exascale and deliver Levi's Key so that the whole system would speed up, all in a matter of minutes. And I saw some of the gear that Kyle was carrying in that bag. Extremely specialised."

I was caught on a spot now. I would need to decide whether to confide in Juliette. And if I did, how much I was going to tell her of our speculation.

I dived in, with a feeling similar to that when the system was switched on, back at the Lab.

"Okay, I'll come clean, I know some people in the UK Secret Service,"

Juliette laughed, "That's like the bit in a movie where someone has to break cover and the partner doesn't believe him!"

"I know, it is awful, but I really do. I built the cyber cash mining system with Kyle and another friend - Tyler- when we were all students together. The one that you found written about in the Brant archives. It attracted unwanted attention from the US NSA and UK's GCHQ and then I was contacted by someone - Amanda- from the UK Secret Service about the whole situation."

"If I didn't know you by now, then I'd think you were spinning me a tale," said Juliette, "But in the interests of sanity I will say that I believe you."

"Well, this next piece doesn't make complete sense, I made friends with Simon Gray here in Geneva and he told me he was from MI5. He then invited me back to London to meet Amanda - again - and she and her colleagues briefed me about something strange happening here in Brant."

"I didn't know Bigsy, Chuck or Christina until I flew back to London during the first weekend of my time in Geneva, which was then I got briefed about all of this. Bigsy, Christina, Jake and someone named Clare all work for an outfit called 'The Triangle' which seems to sub-contract some work on behalf of the UK Government. Amanda trusts them implicitly and they also have a link into GCHQ, which happens to be where someone else that I know works: Grace, from my cyber currency days."

"Okay," said Juliette, "This is a lot to take in. Has anyone found out any more knowledge about Levi, from all of this?"

I thought the next piece would be difficult to say to Juliette.

"There is a theory. It is that Simon was on the yacht with Levi, pushed him off, and then a scuba diver pulled Levi under the surface until he drowned. Then, Simon and the scuba diver were rescued by another boat and made their way to shore."

Juliette guessed, "And the scuba diver was Bérénice, and the other boat was a motor launch being run by Niklaus?"

"How did you know?"

"You may have only been looking for a few days, whereas I've been thinking about this for weeks," explained Juliette, "Simon had so much to do with Levi's boating, and I noticed that Simon and Bérénice seemed close."

"You won't have heard the latest about Simon, though?"

Juliette shook her head, "No, has he been found?"

"No, it's grisly news actually. Simon flew to the USA, changed his identity to Sacha something, but US Border Control found him. The US authorities detained him on an air base, but he escaped. His escape plane crashed into the desert in New Mexico."

"Was he found?" asked Juliette,

I answered, "No and the other co-pilot wasn't found either - although he did eject."

Juliette said, "Ew, this is all a bit of a mess!"

Universal Soldier

He is five feet two, and he's six feet four
He fights with missiles and with spears
He is all of thirty-one, and he's only seventeen
He's been a soldier for a thousand years
He's a Catholic, a Hindu, an atheist, a Jain
A Buddhist and a Baptist and a Jew
And he knows, he shouldn't kill
And he knows he always will
Killing you for me my friend, and me for you

And he's fighting for Canada, he's fighting for France
He's fighting for the U.S.A.
And he's fighting for the Russians
And he's fighting for Japan
And he thinks we put an end to war this way
And he's fighting for democracy
...
He's the one who must decide
Who's to live and who's to die
And he never sees the writing on the wall.
...
He's the one who gives his body as a weapon of the war
And without him all this killing can't go on
He's the universal soldier and he really is to blame
His orders come from far away, no more
They come from here and there, and you and me
And brothers, can't you see?
This is not the way we put the end to war

Songwriters: Buffy Sainte Marie

Word travels fast

Word of the rat experiment travelled fast.

By the next day, Tuesday, Amy returned to the Lab with the news that our entire entourage would be attending a trade fair in Barcelona, Spain. It was euphemistically called the 'Future Intelligence Expo' but most of the exhibitors there were taking about advanced electronic warfare. A few were presenting on civil policing and Brant and Raven were two of the larger exhibitors present.

Kjeld Nikolajsen was to present there on the topic of 'Gazing into the Crystal Ball – What will HCCH look like in the next decade?'

It read: We are living through the information age and conflict management is now a multi-domain effort. There are significant investments in information capabilities, but what does this mean for HCCH (Human to Computer to Computer to Human Interaction)?

Network-centric conflict management was the buzz

word of the last decade, however this revolution may have allowed sensors to speak to combatants and a more informed OODA (Observe, Orient, Decide, Act) loop for the commander, but every platform and soldier was not also a sensor.

Now with the advances in networked platforms, all platforms become highly capable sensors. Why now, what has changed?

Technology comes in a smaller form factor with more processing power and ubiquitous connectivity. This has provided software-augmented offensive capabilities that can be repurposed as sensors, additionally providing offensive cyber capability, all whilst being networked.

This is not the technology of tomorrow, it exists now, and its adoption will change the role, structures and functionality across most conflict theatres.

"My God," said Juliette, "Brant are trying to make the Universal Soldier."

Kyle laughed and said, "Well Brant's PR certainly know how to make a Press Release buzzword-compliant!"

Rolf looked at me, "I guess we'll need to try a few more times with the Cyclone before we go to Barcelona!"

Pointillist episode

The next week and a half passed by quickly. I'd found more of the routine, at the Labs. They tried a few further experiments, and I could feel that the Cyclone and my brain were somehow becoming closer. It really was like eyesight where the brain turns the eye's vision the right way around for it to be useful, and I could tell that my brain was adapting to the signal inputs from the Cyclone.

A couple of nights I even woke up, when I experienced the pointillist effects of the Cyclone filling in the detail around my area of observation. It was like my brain was either housekeeping or trying to figure out what the dots and flashes of coloured light really represented.

It was obvious that Kjeld Nikolajsen wanted to make a splash at the Barcelona Expo. He even visited our lab a couple of times, once with Amy and once by himself.

He seemed desperate to be able to have something 'announcement worthy' and I just hoped that the novelty of what we'd already achieved wouldn't wear off before the Barcelona Future Intelligence Expo. He imposed strict security onto our lab. We didn't tell him that we had added some layers of our own.

Kjeld wanted a demonstration at the Barcelona Expo, but we persuaded him it was too difficult to achieve live and that we should video a session and then project it.

"After all," Rolf explained, "That's how Apple show off its new toys and their shows are always very good."

To our surprise, a team from 'One Smart Cookie' turned up and explained to us how they were going to make 'Industrial Theatre' of our discovery. I immediately had visions of the Rocky Horror Picture show, but they said, no, they could make something far more visionary.

They likened our discovery to ending on another planet and before we knew it, we were in Kubrick-style space fantasy. I could get their point though. We had all the high-tech gear and the twinkly lights, so it wasn't so difficult to envisage the mind-meld as landing on another planet.

Kjeld even said, "In our wildest dreams, we would not have dared to think that our new partner One Smart Cookie would be so vastly superior to our previous partner; the production so surprisingly fresh, the quality of the production team so top rate, the execution so crisp, the direction so articulate."

And then they found the song, although I like to think I had a hand in its choice. Juliette loved it:

"Another girl, another planet,' which featured the lines

I think I'm on another world with you
With you
I'm on another planet with you
With you

Now we had a soundtrack for The Presentation. The future was so bright that Kjeld better wear shades.

We'd rerun the rat connection experiment and even tried a variation where we'd connected two back-packed rats together. That was interesting because they both stalled and then went into a territory marking role. Like marching around the perimeter of the cage. It reawakened my worry about using animals as robotic cyber-beings.

I'd used the Cyclone headgear several times now, over the course of the previous week, and still always had that sensation of falling into a dark chasm as the full Exascale configuration started up.

Then something else started happening. It was something I could remember from my ancient experiments with the tDCS otherwise known as transcranial direct current stimulation. That's when I was hooking up electrodes to my skull and then blipped it with a 9-volt battery.

I'd sometimes get after-effects from that, which somehow told me it was doing something. I was finding similar effect now when I fell asleep. I could see rain showers of coloured lights. They were like the lights I'd see under the Cyclone, but these seemed to come of their own accord. I disturbed Juliette a couple of times, and she woke me up, and I swear that once the lights carried on for several seconds, much like when I was under the influence of Levi's software.

I told Juliette about it, to see if, as a psychologist, she has a psychiatric theory. She came up with some theory and a bunch of three-level models. I'm inherently suspicious of the tidiness of three-level models, but one of them seemed to fit the situation.

- **EPISODIC MEMORY** stores information about when events happened and the relationship between those events and relates to personal experience.

- **SEMANTIC MEMORY** is the organised knowledge about the world. Essentially all the "facts" we have accumulated.

- **PROCEDURAL MEMORY** involves knowing how to "do" something. It relates to skill learning, the connections between stimuli and motor behaviour.

It also implied that my flashing lights memory was bouncing about in the episodic, rather than getting down into the semantic or procedural levels. It was like cat memory. I couldn't count past six and therefore I could not unlock the data contained in the flashes of light.

That's when Juliette had a brainwave. She was talking about the flow state, like when one listens to classical music when studying and more of the information sticks.

She wondered if I'd be better if I could reach a state of mellow.

"Yes, you could try being 'in the zone' - concentrated, outside of reality, inner clarity, serene, timeless, intrinsically motivated," she said.

"Oh No! Stress from so much to remember and to do!" I made that face like on the Edvard Munch painting of The Scream. I remembered Munch has written 'Can only have been painted by a madman', in pencil in the top left-hand corner.

"Okay, it'll be the Mozart, then!" I replied. The next session would be accompanied by music.

MilSpeak

Thursday's working session was the last before we shipped out to Barcelona. The equipment was being shipped out by air freight, and we were to follow in good time to set up the test rig in the conference centre. In other words, we needed to be there on Sunday, a day before the Expo started on Monday late morning.

Hermann had judged that it was too much risk to take our main system, and so he had spent the preceding week making a remarkably good replica, except the it didn't include Levi's Key. It could look the part, would function, but would run as slowly as my first experiment.

He'd embellished it somewhat, too. He'd added an intercept to the Cyclone to Computer link.

"It's really to make it look more impressive, but I'll say we can handle language translation through this interface," he explained.

"I'm taking an audio feed and then sending it to Google translate or to Siri. Then I bounce the translation back the other way. It's not real-time but it gives the participant a sense of context."

We'd use the amazing videos made by 'One Smart Cookie' to provide any further demonstrations and would not have to worry about hackers or theft from the Expo hall.

I'd arranged a strange two leg trip to Barcelona, via London. Like when I flew from Cork, not that much different in price from the expensive Swissair flight, and I could use British Airways lounges everywhere because of my card upgrade.

Friday, in London, I was going to Vauxhall Cross for an update with Amanda and the people from The Triangle.

Hermann, Amy, Juliette, and Rolf were flying direct to Barcelona, but I'd made an excuse about visiting a relative in London. I think they thought it was my ex-girlfriend Heather from Cork. I'd told Juliette my plans and asked her to keep them to herself.

I arrived at SI6 just after midday and asked to meet Amanda Miller. Once again, she escorted me through their fast lane system, and I was whisked away to a high-floor meeting room in the Executive Briefing Centre. I walked in, and there, waiting, were Chuck, Grace, Christina, Bigsy, Jake and Clare. No sign of Kyle though, who I guess had gone back to his day job.

"Kyle can't join us today, but he will be in Barcelona for the Expo," explained Amanda, "He recommended that we get hold of Tyler though, and he is on his way here as we speak."

I hadn't seen Tyler since we divvied up the spoils from the cyber cash experiment. I should really be more careful with my friends. Anyway, just on cue, he entered the room with Jim Cavendish. I was impressed that he'd obviously been in a one-to-one briefing with Jim and was now joining the group.

"Hey Matt!"

"Hey Tyler! - Looking good"

"Dunno what I'm mixed up in, rat-boy!" said Tyler smiling.

I realised this was the English equivalent of the faire la bise. Bantz. Banter.

Jim looked sheepish, and I worked out that Tyler had received a full briefing.

We all sat around a big oblong table and Amanda began to speak,

"Grace thinks she has found something. A deal is being brokered between Brant and the Chinese State. I'm going to turn over to Grace to explain more. I just want to say that what we are hearing about is commercially sensitive and we have obtained it by a rather roundabout route."

Grace began," Yes I cannot emphasise enough that the is commercially sensitive. Some of you might think it is predictable, but it would dramatically boost Brant's share price if the knowledge leaked out."

"Both Brant Industry and Raven Corp's share prices, " added Amanda.

Grace continued, "In the past, nation states have come together to prohibit particularly gruesome or terrifying new weapons. By the mid-20th century, international conventions banned biological and chemical weapons. The community of nations has forbidden the use of blinding-laser technology, too. A robust network of Non-Governmental Organisations has successfully urged the UN to convene member states to agree to a similar ban on killer robots and other weapons that can act on their own, without direct human control, to destroy a target (also known as lethal autonomous weapon systems, or Laws).

"And while there has been debate about the definition of such technology, we can all imagine terrifying weapons that all states should agree never to make or deploy. A drone that gradually heats enemy soldiers to death would violate international conventions against torture; sonic weapons designed to wreck an enemy's hearing, balance or heart rhythm should merit similar treatment. A country that designed and used such weapons should be exiled from the international community.

Amanda said, "In the abstract, we can probably agree that ostracism – and more severe punishment – is also merited for the designers and users of killer robots. The very idea of a machine set loose to slaughter is chilling."

"But, said Bigsy, "I suppose the argument runs that

anyone prepared to build such things isn't that bothered about being ostracised. 'Might is Right,' and all that. Amorality, consequentialism, and psychological hedonism rule. Survival of the fittest..." Bigsy trailed off.

Grace spoke again, "Bigsy make a good point, and some of the world's largest militaries seem to be creeping toward developing such weapons, by pursuing a logic of deterrence: they fear being crushed by rivals' Artificial Intelligence if they can't unleash an equally potent force. It's like another arms race."

Jim spoke, "The key to solving such an intractable arms race may lie less in global treaties than in a cautionary rethinking of what martial AI may be used for. As 'war comes home', deployment of military-grade force within countries is a stark warning to their citizens: whatever technologies of control and destruction you allow your government to buy for use abroad now may well be used against you in the future."

Grace continued, "Then there is the debate...Are killer robots as horrific as biological weapons? 'Not necessarily,' argue some establishment military theorists and computer scientists."

Grace continued, "Now this may shock some of you, but I have here the opening section of Kjeld Nikolajsen's speech at the Expo next Monday."

"Nikolajsen begins:

'According to the US Naval War College, military robots could police the skies to ensure that a slaughter like

Saddam Hussein's killing of Kurds and Marsh Arabs could not happen again. A spokesperson at the Georgia Institute of Technology believes that autonomous weapon systems may 'reduce man's inhumanity to man through technology', since a robot will not be subject to all-too-human fits of anger, sadism or cruelty.

'Michael Schmitt proposed taking humans out of the loop of decisions about targeting, while coding ethical constraints into robots. Ronald Arkin developed target classifications to protect sites such as hospitals and schools.

'In theory, a preference for controlled machine violence rather than unpredictable human violence might seem reasonable. Massacres that take place during war often seem to be rooted in irrational emotion. Yet we often reserve our deepest condemnation not for violence done in the heat of passion, but for the premeditated murderer who coolly planned his attack. The history of warfare offers many examples of more carefully planned massacres. And surely any robotic weapons system is likely to be designed with an override feature, which would be controlled by human operators, subject to all the normal human passions and irrationality.'

Grace paused and then said, "There...That's the opening section from Nikolajsen's speech and I'm sure you can see it is positioning that robot war machines are favourable, and that warfare can be encoded. It's a predictable Brant position, probably spun up by their PR firm."

"Honestly, "I said, "I'd be astounded if Kjeld could say

most of that. Jasmine Summers seems to represent the thought police around at Brant, and I imagine she'd want something glitzier."

Jim spoke now, "Any attempt to code law and ethics into Artificial Intelligence raises enormous practical difficulties. Several scientists have argued that it is impossible to programme a robot warrior with reactions to the infinite array of situations that could arise in the heat of conflict. Like an autonomous car rendered helpless by snow interfering with its sensors, an autonomous weapon system in the fog of war is dangerous."

Chuck nodded, "Most soldiers would testify that the everyday experience of war is long stretches of boredom punctuated by sudden, terrifying spells of violent disorder."

Tyler commented, "Standardising accounts of such incidents, to guide robotic weapons, might be impossible. Machine learning has worked best where there is a massive dataset with clearly understood examples of good and bad, right and wrong. You have to know in which quadrant to place the action."

Tyler continued, "For example, credit card companies have improved fraud detection mechanisms with constant analyses of hundreds of millions of transactions, where false negatives and false positives are easily labelled with nearly 100% accuracy."

Chuck asked, " Would it be possible to 'data-ify' the experiences of soldiers in Iraq, deciding whether to fire at ambiguous enemies? And if it were, how relevant would such a dataset be for occupations of, say, Sudan or Yemen?"

Tyler was shaking his head.

Grace continued, "Chuck is right. Given these difficulties, it is hard to avoid the conclusion that the idea of ethical robotic warriors is unrealistic, and all too likely to support dangerous fantasies of pushbutton wars and guiltless slaughters. I'm afraid that is what Brant's ideas are leading toward."

"Yes," I said, "Kjeld's speech is almost too honest and forthright."

Amanda spoke, "International humanitarian law, which governs armed conflict, poses even more challenges to developers of autonomous weapons. A key ethical principle of warfare has been one of discrimination: requiring attackers to distinguish between combatants and civilians."

Grace spoke now, "But that pond has been muddied. Guerrilla or insurgent warfare has become increasingly common in recent decades, and combatants in such situations rarely wear uniforms, making it harder to distinguish them from civilians. Given the difficulties human soldiers face it's easy to see the even greater risk posed by robotic weapons systems."

Amanda spoke, "That's how we get companies like Brant making such pro-machine statements. As proponents of such weapons, they insist that the machines' powers of discrimination are improving. Even if this is so, it is a massive leap in logic to assume that commanders will use these technological advances to develop principles of discrimination in the din and confusion of war."

Jim added, "The principle of distinguishing between combatants and civilians is only one of many international laws governing warfare. There is also the rule that military operations must be 'proportional'; they must strike a balance between potential harm to civilians and the military advantage that might result from the action."

Chuck added, "Well, the US Air Force has described proportionality as 'an inherently subjective determination to be resolved on a case-by-case basis'.

Tyler laughed.

Chuck looked at him. "Yeah, it's MilSpeak ain't it? No matter how well technology monitors, detects, and neutralises threats, there is no evidence that it can engage in the type of subtle and flexible reasoning essential to the application of even slightly ambiguous laws or norms. When we were flying missiles in the Arizona desert, they were incredibly accurate, but relied upon the human observer's discrimination to know which target to blast."

Grace continued, "They say that an iron fist in the velvet glove of advanced technology such as drones can mete out just enough surveillance to pacify the occupied, while avoiding the devastating bloodshed that would provoke a revolution or international intervention. In such a robotised vision of 'humane domination', war would look more and more like an extraterritorial police action; hundreds or thousands of unmanned aerial vehicles patrolling the skies creating a disturbing fusion of war and policing."

Jake questioned this, "I assume that Nikolajsen is promoting these scenarios as a way to boost Brant's products and share price?"

Tyler commented, "Yes - it looks as if Kjeld Nikolajsen has taken an entirely literal position and plans to use it to present the findings."

Amanda nodded, "Yes, how should global leaders respond to the prospect of these dangerous new weapons technologies? One option is to come together to ban outright certain methods of warfare. It has partially worked in the past. Landmines were banned by most countries, except, notably, the United States. Now, instead of bans on killer robots, the heavily lobbied US military establishment prefers regulation. Concerns about malfunctions, glitches or other unintended consequences from automated weaponry have given rise to a measured discourse of reform around military robotics."

"This is all playing to Brant's position though, isn't it?" asked Tyler, "Invent something, persuade everyone to like it, sell lots and make a happy conscience-free profit."

Chuck nodded now, "Weaponry has always been big business, and an AI arms race promises profits to the tech-savvy and politically well-connected. When I was out testing missiles in the desert, we were interested in their range and accuracy, but not the human commands and actions to deploy them. You had to trust that they were being aimed at the right things."

Grace said, "Counselling against arms races may seem utterly unrealistic. After all, nations are pouring massive resources into military applications of AI, and many citizens don't know or don't care. But that attitude may change over time, as domestic use of AI surveillance ratchets up, and they increasingly identify the technology with shadowy apparatuses of control, rather than democratically accountable local powers."

Grace continued, "It brings us to China. In China, the government has hyped the threat of 'Muslim terrorism' to round up a sizeable percentage of its Uighurs into re-education camps and to intimidate others with constant phone inspections and risk profiling."

Bigsy added, "I've seen those 5G Patrol Robots. Some look like miniature petrol pumps and others are like children's expensive electric toy cars. The Chinese are using them for 'Healthcare' and 'Traffic' measures officially, but they all have facial recognition and phone tracking built in. They brought them in for pandemic management but can just as easily track a perceived problem citizen. Some of them have rudimentary AI to issue warnings and take photos of people - for example in areas where they are not supposed to be, when they should wear masks or jay-walking, or even when citizens are over-temperature and may carry influenza or worse."

Grace added, "And that's what the other part of the

documents I received implies. That Brant is positioned to do a deal with China. The United States could try to block it, like they did with 5G and so many other technologies, but that is much more difficult to do when the base company is based in Switzerland."

Amanda added, "No one should be surprised to know that Chinese equipment powers US domestic intelligence apparatus, while massive US tech firms get co-opted by the Chinese government into Chinese surveillance projects."

"Ker-ching," said Jake.

Grace added, "The advance of AI use in the military, police, prisons and security services is less a rivalry among great powers than a lucrative global project by corporate and government elites to maintain control over restive populations at home and abroad. Once deployed in distant battles and occupations, military methods tend to find a way back to the home front."

Clare cut in, "Well then, I can see that the AI Headgear invention, based on Levi's design, almost works now and is to be launched in Barcelona. This will push the share prices up for both Brant and Raven, based on rather immoral projected uses."

"Well summarised," said Chuck, "So who will be in Barcelona next week? And how do we stop this thing?"

Barcelona Airport

Saturday around midday, I flew to Barcelona from London Heathrow and invited Clare and Bigsy to join me in the Concorde Lounge. It impressed them that I'd got the magic card, but I noticed they both had Gold Cards, anyway. Kyle had somehow gained a similar magic card. I didn't dare ask. Tyler had said he'd be along, but was making his own route, flying from City Airport. Something about needing to meet someone in Amsterdam on the way. He'd certainly have time, with an 8-hour stopover.

The Createl gang would fly from Geneva on their expensive plane to get in on Sunday around midday. They still had enough time to prepare, although Hermann and Rolf had flown Saturday to check that the kit was correctly set up. They had booked into a different hotel close to the water.

I pinged Rolf as soon as we landed, and he replied: 'Hope all good with family ;-) All set up and fine. Hermann and I are hitting Las Ramblas. Sangria time! BTW - a change of plan. Bob Ranzino is replacing Kjeld for the presentation.'

I took it that Rolf and Hermann would be away in an alcoholic haze for the afternoon and evening, but I wondered why Kjeld had been replaced with Bob.

But Barcelona. What an airport. It seemed to be out of proportion to the size of the city, until I remembered that they had built the airport for one of the Olympic Games. Unlike nearly every other airport, it looked finished. Almost that it still had spare capacity as we walked around the largely empty corridors and vast walkways. They must have received a helluva grant!

A sign pronounced in English and Spanish that Barcelona was one of Europe's top ten airports.

The huge sweeping structure was flooded by natural light through its floor-to-ceiling glass walls.

"Wow," said Bigsy, who, like me, had never visited before.

By contrast, Christina switched into speaking Spanish as soon as we arrived and seemed to know her way around. She mentioned that her friend, Roberta, was working in Barcelona and hoped she could see her whilst in town.

The main hall was called the Sky Centre and processed travellers under a wing-like roof, receiving enormous light via its linear skylights. The long central pier and the side piers suggest the shape of an aircraft.

We met the others at the baggage claim and then all

trooped off together to get a couple of taxis. Amanda and Grace were arriving separately. I had a feeling that Amanda was flying in from Paris and that Grace would fly from Bristol.

Jake said, "No wonder they run conferences here. I once spoke to a conference organiser, who pointed out that only Cannes, France was properly suited to the largest conferences in Europe. It is because of the locations of the hotels and the travel times to reach the main hub buildings where the presentations take place. Hence the Cannes Film Festival."

He added, "They had told me about Barcelona, London and Florence as other useful locations, but pointed out that by the time you were in Florence then things could be a long travel distance apart. Curiously, Disneyland Paris was also on the same short list of useful places."

I could understand why the travel times were important. Get all those executives to come to a foreign location, but then keep them corralled on the business at hand rather than the alternative messy street scenes.

Chuck had said the same, "Normally, if I was in Barca for the sea and sun, then I'd stay down by the beach. Maybe in the W, which is a striking hotel right along at one end. But this time we are here for a conference, so we need to be close to the hub.

Chuck added, "We need a powerful hotel. That's why I suggested the Hotel Fairmont Barcelona Rey Juan Carlos I. It's literally next door to the main event, and everyone can walk there. The big-wig organisers and senior guests will no doubt stay in the Rey Carlos too, so we'll be able to keep 'eyes-on'. I guess a few of the SVIPs will be taken somewhere else, but they are bound to be parked in the Rey Carlos before and after their sessions."

I was fascinated by how Chuck seemed to know about conference management. But then, so did Christina and I worked out it must be something to do with them both operating security details.

Clare, Bigsy and Jake were all aching to get out into the sunshine of late afternoon and they had all heard of Las Ramblas as the place to be.

Chuck smiled, "Well, we are here a day early, we could go along for this evening,"

"Yes Dad," replied Clare, playfully nudging him sideways with her whole body.

Las Ramblas

Chuck and Christina led us to Las Ramblas it looked like the throbbing heart of the city, but I was also aware of the tourist trap potential.

Championing its positive attributes, Christina quoted the poet Federico García Lorca and wistfully explained that he once described the avenue as 'the one street in the world I didn't want to end,'

It has a Parisienne vibe, created by the outdoor cafes and colonnades of slender trees, and the playful performance artists and boisterous carnival atmosphere of this 1.2 km stretch of street.

Christina explained, "Just as it joins the city centre and the sea together, Las Ramblas divides the two most central districts of Barcelona's old town, the touristic Gothic Quarter on one side, and the shabby chic area of El Raval on the other. Clare - you'll have to come with me to El Raval during our time here!"

We walked from one end to the other of Las Ramblas to

get ourselves calibrated, but then looked around for a likely place to eat. Christina said, "It's all planned, I have the perfect place - where I can also say hello to an old friend."

And we played 'spot the conference attendee' as we walked along, noting a succession of people (mostly men) in 'business casual' clothing which seemed remarkably uniform and centred around chinos and variation on polo, golf and soccer shirts, frequently they were hovering outside of KFC or McDonalds.

We decided the shirts represented the management hierarchy.

It was during this game that Kyle ran into one of his old buddies. Kyle's friend looked East Asian and I left them to chat for a few minutes as they swapped up-to-date-details.

"Considering we are both security specialists, neither of us had changed phone number or email," said Kyle, "That was John Yarvay (his real name is Xuan Yahui) who is here for the here Expo. He's originally from China, but when we worked together, he had a great knowledge of German banking systems too. One place we worked, the Germans had written all the security procedures manuals - or so we thought - when we opened the binders containing documentation, they were all empty!"

He paused, "I told him we were with Brant and here mob-handed, but he said if we needed a briefing on Chinese security, he was up for it!"

We hurried to catch up with the others who had kept on walking. We were now in an area with a handful of the city's most upmarket hotels and several great places to eat and drink, including the ubiquitous global fast-food chains.

We noticed, as we got closer to the water, that Las Ramblas increasingly resembled a touristic circus with tacky gimmicks and T-shirts for souvenirs, chronic overcrowding, and soulless consumerism. We were all good at spotting the pickpockets who seemed to hover around the area and an unusual supply of ladies appeared as the evening darkened.

We stopped at a friendly looking popular restaurant, and Christina had a word with the host. He grinned and then called a couple of the waiters who pulled some tables together for us.

We were seated in a dappled shady area, just to the side of the 'main drag' and could see all of humanity go by. And I mean it too. We saw the city bustling in full flow and street performers in the shape of demons, gladiators, clowns, ballerinas, flamenco dancers and many more outrageous guises, entertain, terrorise, mimic, beguile and charm passers-by.

Then the strangest thing happened. A fortune teller was working the tables around us. It looked a little like a man dressed as a gorgeous woman pirate. She was telling fortunes with Tarot cards, then turned toward us and Christina said, "Hello Roberta!"

The fortune teller leapt across to Christina and hugged

her. They started speaking Spanish, but it soon turned to English. Bigsy and Jake were sitting with Christina and she started to introduce them both. "These are the self-same people I was meeting that day when I met you at your place at BoxPark! When you said I was from the land of the ice and snow, and that I was a scryer."

Roberta spoke to Jake and Bigsy, "This woman carries some deep powers. I tried to read her Tarot, once, but she turned the tables and was able to tell me my fortune. It's how I came to be here, in Spain, where I have found my true love and my vocation!"

Roberta passed out a card, it was in Spanish and English and advertised a one-person show, like a Pirate adventure 'All on the Spanish Main' / 'todo en el español principal'

By now, it was obvious that Roberta was a male from London, and as he flipped back into female character, I received a direct gaze.

"You, my friend are the person who needs my powers most tonight. You are about to have something happen. Roberta gazed at me and then from the sleeve of an elaborate purple jacket began to pull cards.

- *The Hanged Man:* Martyrdom
- *The Seven of Swords:* Deception and Trickery
- *The Five of Cups:* Mourning and Loss
- *The Eight of wands:* Quick decisions
- *The Tower, reversed:* Disaster avoided

He laid them in a tidy square, with the upside-down Tower below the other four cards.

"I won't do a full reading tonight," said Roberta, "But these are powerful cards."

Roberta looked.

"I can see the story, someone you have known briefly, they have gone away, or died, and you have suffered their loss. But they were like a serpent and have wrapped themselves around you. But you will see them soon and a disaster will be averted."

I racked my brain at this. In the middle of all the Artificial Intelligence, all I could honestly think of was the rat, but I didn't want to say so. Then I wondered if it had something to do with Heather. Maybe she was here with a trade delegation from Cork? I think I looked confused. Either that or the Sangria was lusher than I had realised.

With a judder, I was aware of the others whooping and cheering and I realised that Roberta's little show was a performance, although it seemed very real to me. I wondered if Roberta had also studied hypnotism.

Chuck had reached into his wallet and pulled out a €10 banknote, Jake and Bigsy did the same and I thought I should as well.

Roberta placed her pirate hat in the middle of the table, and we all tipped the money in. With a small flourish, she picked up the hat and placed it upon her head, with a small flash of sparkles falling from it. She took it off, looked inside, puzzled, and then replaced it.

"Don't be a stranger!" she whispered to Christina, "See you at the BoxPark," and with a flourish of her tailcoat Roberta was gone.

I was sitting between Chuck and Christina, and I realised they knew one another quite well and appeared to have been involved in some prior scrapes together.

Chuck murmured to Christina, "So, are you carrying?"

"I might be," smiled Christina, "but nothing Mulberry-sized."

"It's just useful to know in case we get into any awkward situations."

"I guess there will be so much military-grade hardware at the show, anyway," said Christina.

"So what did you bring?" asked Chuck, intrigued.

"A mosquito," replied Christina, "small enough to disassemble and lose in the luggage."

"You like your SIG Sauers," replied Chuck.

"This is also quite a ladylike pistol," replied Christina, "It's a scaled down P228. I've got it in the same sandy colour as the MPX Copperhead."

"Mmm, tasteful," said Chuck, "But some people say they have a habit of jamming,"

"People who don't look after them, and besides, I swapped out the magazines for .22 micro mags. They don't spoil the line of what I'm wearing."

Chuck laughed.

"I suppose you have brought a big shooter?" smiled Christina,

"Yep, and I had the paperwork to accompany it - EFC - so I guess I'm more legal than you are," said Chuck.

I worked out that Chuck was referring to a European Firearms Certificate.

"So, what is it?" asked Christina, intrigued.

"A P226," answered Chuck.

"What? - That's a handgun, hardly a hunting rifle."

"No, I brought it in with the pistol carbine conversion kit, all locked up in a business-like black gun box."

Christina looked genuinely surprised, "But that would have made it look like something that a Star Wars Space Trooper would carry."

"I know, hidden in plain sight. And all properly documented too. Now I can do menacing or tactical lightweight, but the Spanish think I'm here for the grouse."

I had no idea what they were both talking about but imagined that Christina's pistol would look like a colourful toy, and Chuck's like something to be used at Armageddon.

Sometimes it happens sooner that you'd think

The Boqueria's food was amazing. We had a vegetable salad tapas selection, then the paella, with clams, crevettes, prawns, and mussels, and the patatas bravas. To round things off the crema Catalana. It was a treat to unwind, and I felt very relaxed in this company, despite the curious remarks from Roberta.

Then it was time to move on. I decided to go back to the hotel, because I'd have to meet Rolf and Hermann in the morning and the others from Brant later in the day. I also received a message from Tyler to say he was now in Barcelona and whatever he'd been doing in Amsterdam was successful.

Clare decided to duck out too, so we headed for the main road at the end of Las Ramblas and hailed a cab back to the hotel, leaving the others to go onwards for a wild night.

I could tell we'd wimped out, because when Clare and I returned to the hotel, there were other delegates still getting ready to go out for a late night in Barca.

I said goodnight to Clare with an old-fashioned hug, and we were just about to go our separate ways, when I saw a familiar face in the check-in line.

Simon Gray. Very much alive.

I whispered into Clare's ear and she turned to look. I'd forgotten that Clare had got up close and personal with Simon when we both visited London for that first briefing.

"She nodded and pushed me around a corner to where there was a row of hotel phones on the wall, and there we waited until we could see what he was doing.

He checked in, and was now making his way to his room, oblivious to us watching him. An advantage for us was the American design of the hotel, which had a huge central atrium, with all the hotel rooms around the edges. We could see his glass-sided lift go up to the fourth floor and see him get out. We watched as he found his room, three across from the elevator.

Suddenly the evening's Sangria had worn off and I checked with Clare.

"We are sure that was him?" I asked her again. I wanted to make sure I was not under some dodgy hypnosis from Roberta.

"It was definitely him," asserted Clare, "So much for being killed in a plane crash."

I said to Clare, "I'm going to call Chuck, maybe you could also call Christina?"

We both picked up our cellphones and dialled, Chuck picked up and was in a noisy night club. I waited until he moved to somewhere less busy.

"I've just seen Simon Gray; he's here, in Barcelona, staying in our hotel. Clare agrees it's him."

"He's dangerous," answered Chuck, "Don't do anything until I get there. It'll take 20 minutes."

Clare had just had a similar conversation with Christina, although her estimate to return was fifteen minutes. I wondered if the group had stayed together.

Sure enough 15 minutes later, a taxi arrived with Chuck, Christina, and Jake. They spotted us in the foyer and told us that the others were following back in a second cab. Jake had also called Amanda, who had, by now, arrived and was in her hotel room.

"We have to grab Simon now," said Chuck, "before he has a chance to do anything."

Christina nodded, "We should go armed."

Tainted Love

Sometimes I feel I've got to
..Run away
I've got to
..Get away
From the pain you drive into the heart of me

And the love we share
Seems to go nowhere
I've lost my light
For I toss and turn I can't sleep at night
Once I ran to you (I ran)
Now I run from you

This tainted love you've given
I give you all a boy could give you
Take my tears and that's not nearly all
Tainted love
(Ooh) Tainted love

Songwriters: Edward C. Cobb

Disruptive Pattern Material

By now we had moved across to the seating area which covered a large part of the Hotel's lobby. We would wait for Amanda before doing anything. Chuck pointed out that the despite its name, 'Future Intelligence Expo' was primarily militaristic in nature and that a good proportion of the attendees would be from varied armed forces around the world.

We noticed that several of the delegates were in military clothing, although others looked like they were in 'off duty soldier' clothes, casually jacketed and booted. Chuck and Christina were playing spot the nationalities, based on the type of uniform. I learned that ACU (Army Combat Uniform) was different from MARPAT (Marines Pattern) and that they were both different from DPM (Disruptive Pattern Material), which was worn by the Brits. A few soldiers from more rigid regimes were in 'walking out dress' which seemed to be more or less standard uniform, at least to the untrained eye.

Kyle commented, "I haven't seen this much uniform at a civilian event since I was in Tel Aviv, when half the delegates came dressed in their uniforms. And that was for a general IT Expo. Look at those two - what are they? Chinese?"

A couple of Chinese police - a woman and a man in full black uniform were standing in the middle of the foyer. They were both wearing dark glasses.

"Do you see, those two?" asked Kyle, "They are both using surveillance glasses. Those dark glasses have a camera built into the edge, a bit like the old Google glass. They are recording the delegates and no-doubt collecting them into a database."

"Oh - that's so creepy and it's all in plain sight too."

"Have you been to Beijing? It's got more cameras than London!" said Kyle," And not only that, but they also all work and are tied back to a ginormous database of images."

Just at that moment, we saw Amanda Miller. She was walking towards the check-in area and Chuck hastily stood and walked towards her. They greeted in a surprisingly friendly way and then Chuck turned to point Amanda towards where Jake, Clare, Bigsy and I were seated.

"Oh, wow! The gang's all in town!" she called, smiling, "Even some that we didn't expect!"

Chuck had explained to Amanda about Simon's presence, and that there was a need to apprehend him. I was starting to get rattled by all of these developments and could feel the tension increasing as Chuck and Amanda discussed their next action.

Amanda said, "Of course, I have some backup here, from Britain's finest, but I'd rather not need to call upon them because the whole thing will suddenly become a noticeable high-profile event."

At that moment, a dark-haired member of hotel staff approached us. I wondered if we'd been making too much noise; she didn't look like a waitress, more like one of the management. Then I realised it was Christina, wearing a brunette long-haired wig.

"Are you ready?" she asked Chuck, "We should take him in the room, where we can contain things and it will also be quiet. No need to draw attention to ourselves, particularly because the event hasn't even started yet."

Chuck nodded agreement.

"The rest of you should stay downstairs, Christina and I will be able to handle this," said Chuck.

Christina asked him," You've got the P226?"

"Oh yes," said Chuck, "I had it back in the restaurant, actually."

"Me too, with the mosquito," said Christina.

"Time to scratch that itch," said Chuck and moved toward the glass-sided elevator, with Christina alongside.

"They know what they are doing," said Amanda, "and it will be low key."

A few minutes later, we saw Chuck come to the balcony near Simon's room. He signalled that we could go upstairs.

Amanda, myself and Clare took the elevator. Amanda had picked us as the two who knew Simon the best.

Chuck showed us into the room, where we could see Christina and Simon chatting quietly in Russian. As we approached they switched to English.

"Simon?" I said, "What the hell?"

"I know," he said, "I knew you wouldn't understand, but extra points, Matt, for getting the US Marines and the FSB involved."

I could see a smirk from Chuck at the comment about the Marines.

Simon looked toward Amanda, "Sorry Amanda, but this is all transactional."

I noticed that Christina had a pistol drawn, and it didn't look all that small to me, despite what she and Chuck had said back at the cafe.

"Two ways," said Amanda, "Discreet and painless or noisy and painful?"

"I'll come quietly," said Simon, "As a matter of fact, it would be better if we could make my check-in disappear as if I've never been here."

"We can do that," said Amanda, "And we'll be taking you along the Diagonal to the British Consulate - I assume you are on British papers at the moment?"

She was referring to one of Barcelona's main arterial roads, which the hotel and the Consulate were both located along, barely a couple of kilometres apart.

Simona answered Amanda's question, "Yes, I'm Simon Green, on a British Passport issued at the Britisches Honorarkonsulat in Frankfurt-am-Main. I thought it would take a little longer before it was discovered. I only needed it for one job here in Barcelona."

Christina turned, "Simon must be at least a triple agent. He's Russian, pretends to be a Brit, but was an American in Arizona. I'm impressed."

Chuck disagreed, "No, Simon is a soldier of fortune. A Merc. Mercenary."

Simon pulled a face, "I prefer freelance consultant," he said, "Amanda, I'll be seeking immunity from all of this, especially if I tell you anything new."

Christina spoke, "Simon has already told me he knew about the boating incident but says he didn't do it. He claims to be purely logistic support. Providing the control equipment - radios, monitoring – tech logistics. He says Niklaus worked with another Russian and with

Bérénice Charbonnier but that they had been trying to get information from Levi. Something about a program key?"

I was surprised that Simon had folded so quickly, but I realised that a combined assault on his room by various international security forces would probably rattle most people.

Amanda spoke, "Так как же ты научишься говорить по-русски? Tak kak zhe ty nauchish'sya govorit' po-russki? So how do you know to speak Russian?"

I noticed Christina smile. I assumed that Amanda's Russian was adequate rather than perfect.

Simon answered in English, "My real name is Sergey (Sacha) Sachonovich and I'm from Gorynych in Moscow. The area is run by the Uzbek Bratva and that's who was controlling me. It would be much better if they didn't think I'd ever arrived at the hotel."

"Yes, this way, if I tell you things, it will save us all a lot of time, where you lock me up, feed me gruel, keep me awake and then keep sending in people to ask me questions until I give in. Let's just hit fast forward!"

Amanda continued, " I want to move this conversation to the Consulate. Your passport shows you as a British citizen. You will be in less danger on the inside of our Consulate. I will make the arrangements."

Amanda stepped out of the room and I realised she was arranging for Simon to be taken along the Diagonal to the

Consulate. She returned.

"The car will be here in ten minutes. Grace is going to meet us at the Consulate."

"How will you get me out of the building?" asked Simon.

Christina looked around. She was still dressed as hotel management but relinquished her brunette wig. "Chuck - your coat too, please."

"Come with me," she uncocked her pistol, pulled Simon towards the bathroom and closed the door.

Simon re-appeared a few minutes later. He'd been given shoulder length dark hair by chopping shorter the wig that Christina had brought and a pair of Ray-Bans. It wasn't great but it was an amazing transformation in a few minutes. Then he put on Chuck's bulky coat and I wouldn't recognise him.

'Now we need to get out of the hotel without being recognised, especially not from those Chinese police standing in the lobby." said Amanda.

I was impressed that Amanda had noticed the Chinese police and their penchant for electronic news gathering.

Chuck looked around the room. He picked up the silver hospitality tray which had the coffee cups, sugar and those little milk capsules on it.

"I'll bet you didn't know I could be a waiter," he said.

Christina grinned; she already knew what Chuck was about to do.

Consulate

We texted the others waiting downstairs and they sidled towards the exit. There was a steady stream of taxis and they took one to get to the British Consulate.

Now it was our turn. We all fitted into one elevator but split up when we got to the ground floor. I'd thought we would split two ways, but then I saw Chuck take off by himself carrying the tray and the cups.

A few seconds later there was a crash, and the tray went flying into the air, with the two china cups smashing to the floor. Everyone in the lobby looked around, but by that time Chuck was ambling toward the exit by himself.

I looked back toward Amanda, Christina and Simon and could see that they were already climbing into a black Range Rover. I walked with Clare toward the big revolving door to the outside and together we plucked a taxi from the line, asking to be taken to the British Consulate at Diagonal 477.

As we both slumped into the back of the taxi, Clare

grabbed my hand and squeezed it. I'd never felt more relieved.

Then along to a plate glass block, which said 'ING Bank', but according to both our phones was also the location of the Consulate.

"Pasa por el pasaje para llegar a la Embajada - Pasaje British Embajada" - I worked out that the taxi driver was telling us to go through the passage to find the entrance to the Consulate, or as he was calling it, Embassy. I could see through the passage and noticed several shadowy figures, which I realised were Bigsy, Kyle and Jake.

Clare and I re-joined the others and we made our way inside.

The security inside was low key but asked to see identity. Clare also told them we were with Amanda Miller and the guard nodded. "Are any of you carrying weapons?" asked the security man. I realised that Christina and Amanda must have arrived already. He pointed to an airport scanner system.

At that moment Chuck appeared. He'd heard the question and said, "I am, but you can hold on to it for me." He gingerly produced his P228 and held the handle pointing downwards, between finger and thumb, like it was something slightly unpleasant that he'd picked up from the beach.

The security man didn't seem phased and asked Chuck to remove the magazine. It was only then that I noticed another couple of rather more intimidating guards had appeared from behind a one way-mirror room.

Chuck did as he was requested, ejected a bullet from the chamber and handed all of the pieces to the security man via the tipping safe deposit box system used by banks. The guard passed Chuck a key with a number and walked into a different back room, while the other guards looked on.

By now, Amanda had reappeared from behind the security screens and beckoned us all forward.

"I've requested a meeting room to question Simon, but the British Consulate have insisted that we have a couple of soldiers to guard the room. They are armed, you should know. "

"Oh, and Grace has arrived, she is here as well. We have a very full house."

"And welcome to the British Consulate."

I wondered if the normally quiet Consulate dealing with lost passports and family emergencies was able to handle the amount of disruption we brought.

Amanda led us up to a spartanly furnished meeting room. It had a couple of 'España Barcelona' posters showing the cathedral and la Sagrada Familia - both famous Barcelona buildings by the artist Gaudi. Aside from that, it had one of those Nespresso machines and a large TV.

Simon looked first to Clare, and I saw her smile back.

Amanda picked up the questioning, "Okay, Simon. Or Sergey or Sacha. You've told us you are really a Russian from Gorynych and being run by the Uzbek Bratva. You said you'd 'Fast Forward' and in return we have extricated you from the hotel, brought you into safe custody at the British Consulate where you are under armed protection."

Simon began calmly and not appearing to be under any stress, "Clare, I guess you worked out my tattoos? That's how I got into all of this. Uzbek Bratva got me put in prison and inked me, and then got me released if I'd do some things for them. They knew I spoke damn good English and that I was a scientist. I had to learn about the monitoring gadgets, but it was simple, especially once I'd got a supply of keys to the other apartments. It was so simple to put a key card copier over the original reader in the apartment lobby. I even made it look like an upgrade."

"So did you really learn your English from your mother?" asked Clare, "You told me that your mother was from Chelsea."

I noticed Amanda let Clare ask the questions. An interesting technique, but then Clare and Simon had been together for that evening in London.

Simon answered, "And so she was; she happened to fall in love with my father who was an official in a Russian mining company and they came back to Moscow, although she always had enough money to be able to stay in our apartment in London whenever she wanted."

We kept quiet at this news. Oligarchs with homes in London was, no doubt, flickering through everyone's minds.

Grace stood quietly in the background, like she was assessing the scene.

"But how did you end up in prison?" asked Amanda.

"I was a stupid, stupid young man with a lot of money, and I got busted with an awful lot of cocaine, in my apartment in Moscow. They gave me a five-year sentence."

"That's about typical in Moscow, isn't it?" asked Amanda.

"It is. I honestly don't know whether I was set up, it was a messy time for me and my girlfriend, Tanya, and I can't remember where I'd have got a quarter key of coke from, nor why I'd have hidden it in a kitchen cupboard. But in those days I was often out of it."

"What was her full name?" asked Amanda.

"Oh, Yakimova Tatiana (Tanya) Fyodorovna," answered Simon, "She was also from Moscow, from Kutuzovskaya, near the Third Ring. I met her in a club."

I could see Christina smiling at this. She said, "I think you experienced *Lovushka dlya meda,*"she said, "A honey trap - the Bratva wanted you, so they sent a woman to snare you - it doesn't just happen to foreign businessmen. It's Academy 101. Trap, Release, Threaten, Load Up. -

neschast'ye - Bad Luck!"

"Yes, Christina, You have just confirmed my own thoughts and why I never heard from Tanya again."

Christina said, "We were all told about State School 4, the Soviet Union sexpionage school in Kazan, Tatarstan, southeast of Moscow, on the banks of the Volga river. The school trained female agents to be 'swallows'. It's moved on now, and Russia uses independent contractors as honey traps. "I expect that cocaine was a payment for Tanya."

"Anyway, I came out of prison after a really short time - two months but was indebted to the Bratski Krug - that's like a 'Brothers' Circle' - a high-end Bratva - like a Mafia - who wanted me to do things for them to pay off my debt of getting released. My handler told me it would take three years to pay it off, but it was still much better than being in prison for five years."

"That's when they gave me a couple of small jobs - listening in to a conversation in English - then bugging someone, then having to chat up a scientist in a bar in Moscow. They realised after that gig that I was a knowledgeable scientist in my own right and that's when they asked me to go to Geneva and to set up shop at the apartments."

"So how many of the apartments are bugged?" I asked.

"Oh, pretty much all the refurbished ones, which house new temporary Brant staff. To be honest, I thought it was a waste of time and I can honestly say I never got anything interesting from anyone."

"I was supposed to monitor anyone with a science flair, and I was to look out for the progress being made on the AI systems at Brant. The thing is, people don't talk about their work when they get home. They may make the odd work phone call, but that's about it. The rest is computer tapping, cooking sounds music and the occasional bump or crash. I mainly learned that nearly everyone watches the same TV shows."

"But were you supposed to be spying at the Lab too?" asked Grace, who Simon would have only seen from the meeting in London.

Simon answered, "Truth be told, they realised that I was the wrong type of scientist for what they needed, but they kept me on anyway. That's when, by a stroke of luck, Matt came along. He was in another lab and working directly with the AI. And his apartment was already comprehensively bugged - sorry Mate."

"But back up, Simon, Levi's death had already occurred when I arrived in Geneva?" I stated.

"Yes, that brings in a whole other group of players. Bérénice Charbonnier was already living in the block. I knew she had a colourful reputation, but I was footloose in Geneva and thought, 'what the heck.' We soon became - er - best friends and were in and out of each other's apartments all the time. She knew I was friends with Levi and used to ask me questions about him. I didn't think much about it, because I thought she was simply trying to show an interest in my passion for sailing."

"Then, one day she asked me when Levi would next be out on the lake because her friend Niklaus worked at a boatyard and could arrange to say 'Hi'. "

"I arranged a first meeting between the two of them, and Levi was quite excited because Niklaus had access to a wide range of boats. I left the two of them to it, because as much as I'd enjoyed crewing on Levi's yacht, it was quite stressful. He had a fairly militaristic way to be the Captain, and I didn't always enjoy it."

"Then I heard that Levi was feared drowned and that his yacht was adrift in the middle of Lac Leman. I knew it was a day that Bérénice had planned to go scuba diving, but I didn't realise the connection. I didn't really think about it until the news that Levi had been found in Evian, France on the south side of the Lake."

"Then I realised that Bérénice and Niklaus could have been involved. It was something that I spotted on Bérénice's iPhone messages about a Scuba Party from Niklaus - you know, the ones that just pop up on the locked screen, and when I was around at her apartment, I noticed a couple of yellow marine walkie-talkies. I started to join the dots.

I assumed that Bérénice and Niklaus had somehow pushed Levi from the boat, that he'd drowned and that Bérénice and Niklaus had made their escape in a powerboat. How can I put this politely, but Bérénice has very powerful arms, like she works out and power lifts?"

"I'd noticed that too," said Christina, "That she was very

muscle toned. I guess she could immobilise a medium-sized male single handed?"

"But surely Bérénice couldn't be on the yacht and in the water at the same time?" asked Amanda.

"My thoughts as well," replied Simon, "And that's how I worked out that there must have been another person involved. That way we get Levi and another on the yacht, Bérénice in the water and Niklaus on the other boat."

"It's also part of the reason that I left suddenly for the USA," added Simon, "I suddenly worked it out and that put me in danger."

"The week before Levi drowned, I'd seen Bérénice and Niklaus sitting in a cafe with one of my co-workers at Brant. Her name is Qiu Zheng, and she is Chinese. She works in my Department but is several grades higher, a Department Head who reports directly to MD Mary Ranzino."

Simon continued, "Now my theory - I knew Mary Ranzino worked in China for a couple of years, and I made a link. If Qiu Zhang was under Mary's direction and could control Bérénice - who was very much coin operated - then I could see the beginnings of a conspiracy. Add in Niklaus as an accomplice and you've got a whole story. Except for why...?"

Grace began, "Yes, I was also looking at Mary Ranzino. She was a Massachusetts professor loaned to the Chinese for a couple of years. She was given around $1 million grant money to work in China, from the Chinese government, whilst still attached to a US University."

"So, she was on two payrolls?" asked Clare.

"Definitely, plus the actual grant money," acknowledged Grace.

Grace continued, "There was a lot of talk at the time about industrial espionage as Chinese people tried to work in the US but then carry their findings - largely Intellectual Property - back to China. An example was a researcher, who was arrested at Boston Logan International Airport with 21 vials of biological samples in his bag. Prosecutors alleged he was planning to return to China to continue his research there. Another Chinese robotics worker in the same facility concealed that she was in the Chinese People's Liberation Army."

"So, we get Chinese infiltration of US Universities?" asked Chuck.

"And Research facilities, and for that matter commercial companies," added Grace.

Grace continued, " You may have heard of the Chinese 'Thousand Talents Plan,' designed to keep high-end talent at home, in order to prevent a brain-drain. The country has been losing talent to places like the US and the UK, where hundreds of thousands of Chinese attend top universities and subsequently settle down."

I hadn't head of The Thousand Talents and the blank faces around the room suggested that no-one else had either.

Grace added, "It's the self-same scheme that Mary Ranzino was recruited into, as a simple high-profile acquisition to ensure high quality tuition at the Peking University. Keep Chinese students happy in China with world-class education."

Chuck spoke, "But the US view is that China is repeating a notorious tactic in its development playbook: intellectual property theft. For decades, Washington has accused Beijing of stealing science and technology from the US to gain a competitive advantage. Just like all those copy iPhones. Heck, they even copy the Apple Stores, so some areas are kitted out with multiple fake stores."

Grace agreed, "Yes, the FBI warns that the Thousand Talents Plan could be used by Beijing as a channel to conduct non-traditional espionage, though many reported cases are not related to spying, but violations of ethics, such as not fully disclosing financial conflicts of interest. One could say it is the start of China's version of oligarch-style corruption, learned from the Russians."

Chuck added, "I can remember, the Chinese state tabloids labelled the American scepticism as 'hysteria'."

Grace added, "But Washington increased its scrutiny on China's Thousand Talents Plan, ever since the two countries started to be locked in a trade battle, and Beijing has reportedly refrained from talking publicly about the program. And if we look at the statistics, more than 7,000 researchers and scientists based outside of China have participated in the Thousand Talents Plan, many of whom are of Chinese descent."

Amanda cut in, "But we need to frame this in the current context."

Christina spoke, "There's such an obvious Russian-style game-play in here: Qiu Zhang works for the Chinese MSS. She controls Mary Ranzino and handles Bérénice and Niklaus. They are tasked to get something from Levi. It goes wrong and ends badly for Levi. Like a couple of wolves, Bérénice and Niklaus cover their tracks."

"The MSS?" queried Clare, "That's a new one on me,"

Kyle interjected, "Ah yes. The Ministry of State Security (MSS), or Guoanbu is the civilian intelligence, security and secret police agency of the People's Republic of China, responsible for counterintelligence, foreign intelligence and political security. Its military counterpart is the Intelligence Bureau of the Joint Staff. MSS has been described as one of the most secretive intelligence organisations in the world. It is headquartered in Beijing."

Grace added, "That's right Kyle, MSS facilities operate in the northwest of Beijing in an area called Xiyuan 'Western Park' next to the Summer Palace in Haidian District. It is divided into secretive Bureaus, each assigned to a division with a broad directive. Number 17 is the Enterprises Division - which is responsible for the operation and management of MSS owned front companies, enterprises, and other institutions.

"It's so much like what happened in Russia," said Christina, "When the state apparatus created all of those offshore companies, sent trusted people to run them, launder through them and then appoint other people to take over the state banks and heavy industries like oil, minerals and logistics. The old guard ran everything until the young guns wanted more of the pie and started throwing the old men off balconies."

"I have a feeling that my instincts were right, then," said Simon, "That Bérénice was up to no good, and probably in the pay of the Chinese, orchestrated by Qiu Zhang. I assume it's a grab towards the AI technology by Mary Ranzino, which is surprising because she is already the MD of Brant Createl."

"Simon, I can't understand why you would come back to Europe though?" asked Clare, "After all, you'd successfully disappeared once in Arizona?"

"It was because of the Russians. They planned and ran my escape from Luke AFB. They used fake FBI to get me outside and then used a decoy to apparently fly me away in a jet fighter. It looked like a great escape, but I was still in a hotel just outside the Air Force Base. The two FBI were embedded GRU working for the Bratva Krug. They said my side of the deal was to come to Barcelona and to lure Mary Ranzino to a meeting, on the promise that I was selling the key to the Artificial Intelligence system."

"What key?" asked Clare.

Simon replied, "The one that makes the AI system run properly. With it, the whole Brant research plans leap forward by probably ten years. It makes the company suddenly very profitable."

I had to keep a straight face, I was trying to remember who in the room even knew about the key, which Levi had given to Juliette hidden in that necklace.

I remembered that Bigsy had mentioned it to everyone, but not the significance. We had kept it a secret in the Lab, so only Amy, Schmiddi, Rolf, Juliette and I knew what had happened. I suspected that Bérénice might have suspicions though, because she and Niklaus had been sent after Levi.

"Do you think Bérénice and Niklaus were trying to get a key from Levi? And then the plan went wrong?" I asked.

Simon answered, " That's about the shape of it, and it's being sponsored by the Chinese."

The Wolf covers its Tracks

Oh my name it means nothing,
And my age it means less.
I'm far from my home,
High above the mid west.

And as I look down,
On some far city lights,
I pray God to look over,
My family tonight.

When death comes without warning,
With God on its side.
When they call on their guard,
To justify their attacks,
Just serves to remind me,
The wolf covers its tracks.

-   Stephen William Bragg

Expo

Next morning and it was time to walk in bright sunshine from the Hotel across the way to the Palau de Congressos de Catalunya, which was where the Expo was running.

I'd messaged each of my Brant colleagues to say I'd been caught up in some business with Amanda and Chuck and received a variety of replies from "Hope everything is okay?" (Juliette) to "What happens in Barca stays in Barca" (Hermann).

I'd arranged to meet Tyler for breakfast and then to get to the venue early and to meet Amy, Hermann, Rolf, and Juliette at the impressive Brant stand.

Breakfast first and we were both eating scrambled eggs and bacon (but not as we'd know it).

"So come on, Tyler, what was the side trip to Amsterdam about?"

'Remember Erica?"

"How could I forget Erica?" I asked. Erica had been Tyler's long-legged girlfriend when we shared a flat in London. They'd broken up, but he still seemed to care for her.

"Well, after that business with Drew, Erica decided to go her separate way. Erica moved with her Bank to Amsterdam, which was an up-and-coming financial centre, targeting to acquire business post-Brexit. There, she met Evert-Jan Fischer, who was a researcher at ICAI."

"ICAI?" I queried; I'd never heard of it.

"The Netherlands Innovation Center for Artificial Intelligence (ICAI). ICAI has been involved as one of the leading scientific AI communities with research labs throughout the country. They have been working on language interfaces.

I thought I'd ask Evert-Jan they had anything which could help Rolf with his AI Language interface, and they came up - almost straight away - with this little circuit board in a box. It's pretty much a plug and play device and I'm sure Rolf will know how to connect it to the rig."

"Whoa - though. Wait a minute. How are you still in contact with Erica after all this time? And how come her boyfriend will be prepared to meet you?"

"I know, it kinda sucks doesn't it? It was Erica who contacted me when she first moved to Amsterdam. She knew I'd worked there and relied upon me to get her sorted out - more than her bank even - But I realised I'd really moved on - there was nothing except platonic help available from my emotional reserve. I decided I'd tell her about Rosie, and that should finish everything.

"Rosie?" I asked. I decided it would have been simpler if I'd kept notes.

"Yeah, you remember Rosie. Rosemary Marr? I was lightly dating her back in the day?" I suddenly remembered Rosie as the bright woman from Tyler's old office.

"Mate, I had no idea. Didn't she work at your office?"

"Yes, and we fell for one another."

"I bet that was complicated?" I asked.

"You have no idea. But I left working for that crowd after we'd sorted out the cyber coin thing. It streamlined things somewhat!"

I looked across the table and spotted Tyler's watch.

"Man, we are almost late! We'd better go hit the Expo."

We sprinted into the sunshine and across the road to the Exhibition centre.

Our exhibit was there, including the headgear and several monitors playing looped videos of the experiments. I've got to say it all looked slick.

I wouldn't tell any of them about yesterday evening's fun and games, and it was obvious from Rolf and Hermann's looks that they had also had a fairly extended evening in Las Ramblas.

They greeted Tyler when we arrived, and Rolf and Hermann immediately started asking him questions. I saw Tyler passing Hermann a package which I assume contained the specialised circuit board.

Juliette came over to check with me. She whispered, "Is everything okay?"

"It's fine, we had to do a little containment yesterday evening. It's best that I don't tell you too much."

She looked at me, "I knew it," she said, "He's here, isn't he?"

"Who?" I asked.

"Simon. I thought I saw him in the lobby yesterday evening. It was him, wasn't it?"

"Please don't say anything, but yes, it was."

"How did he get away? From the plane crash?"

"He wasn't on it. It was a set-up. He is working for the Russians."

"Is he still here? Running around?"

"No, he is under lock and key - I won't say where though."

Hermann joined us, "Well, we've got a very presentable fake system here," he said. He gestured toward the Exascale computer and the Cyclone headgear.

"It all works in PowerPoint!" he said as gestured expansively at all the equipment on display in the Brant area.

We turned to look, just as Bob Ranzino and the woman from HR who had been present at my hiring interview walked towards us. Bob seemed to have an even deeper suntan.

"Hey everyone, good job," said Bob, smiling, "And I asked Jasmine here to give my presentation the once-over,"

Jasmine smiled, "It didn't really need many changes, just a few more mentions of Brant, and the great work we are doing in Healthcare and Government."

The main doors to the event were not due to open until 10am, so there was still over an hour to go but Bob said he had to go and get miked ready for the keynote session.

"Drive-by handshake," muttered Hermann to Rolf when he was sure they were both out of earshot. Amy heard, though, and grinned.

"Are we all set?" she asked, and Hermann nodded, "Yes, all the screens work, the demo replay is functional and our badge scanners are functioning. The appointments system is working so that we can run specific sessions here in the breakout area as well as back in the VIP briefing suite at the hotel."

Hermann had got everything covered, and there would be enough Brant execs on hand to deal with high-end clients. Our job was to look scientific, down here in the demo space.

"And look, I've got a feed from their internal television, we can put it onto any of the monitors. We'll be able to see Bob's pitch," announced Rolf, triumphantly.

Amy was scouting our near neighbours' stands, to see if there was any adjacent obvious competition for Brant (there wasn't) and to see whether any of the stands would be likely to pull in crowds.

Predictably, behind our stand was a loud stand with dramatic golf and football screens where the delegates could try to score virtual goals or drive virtual holes-in-one. We knew it would bring them large numbers of people, but that they would not really be interested in the products being displayed.

PART THREE

See my train a comin'

Well, I wait around the train station
Waitin' for that train
Waitin' for the train, yeah
Take me home, yeah
From this lonesome place
Well, now a while lotta people put me down a lotta changes
My girl had called me a disgrace

Gonna leave this town, yeah
Gonna leave this town

Gonna make a whole lotta money
Gonna be big, yeah
Gonna be big, yeah

I'm gonna buy this town
I'm gonna buy this town
An' put it all in my shoe
Might even give a piece to you

That's what I'm gonna do,
What I'm gonna do,
What I'm gonna do

Jimi Hendrix

Bob's keynote

Bob Ranzino had been given a keynote spot at the conference. We could see it on the monitor feed that Rolf had created. There had been a rush of delegates who passed through our Expo space but were on their way to the main presentation. No one loitered to look at our stuff, nor at the golf behind us.

"It'll be better here in the coffee breaks and tonight when we offer the delegates some free wine," said Hermann.

Chuck, Christina, Clare, and Jake had gone through to the conference auditorium for the keynote speech. Bigsy and Kyle remained in the demo area with Rolf, Hermann, Amy and Juliette.

Bob started his speech about Barcelona Future Intelligence Expo and managed to weave in a few words from Federico Garcia Lorca. He was being questioned by two well-known American TV anchor-persons Dan Anderson and Rachel Jackson, supposedly to bring a balanced viewpoint.

"Every step we take on earth brings us to a new world," he quoted. It was Lorca.

I gasped when I heard this, which seemed to be taking Lorca's body of work and then extracting one sad counterpoint from it.

Lorca was murdered by a rightwing firing squad, but it rekindled in me Lorca's aphorism that 'a dead man in Spain is more alive than a dead man anywhere else in the world.' Bob had clearly been given a pitch-start quotation and an accompanying graphic that was so schmaltzy that it would do well on one of those inspirational calendars.

Lorca's murder arguably made him the most famous martyr of the Spanish civil war and a symbol of the rampant anti-intellectualism and intolerance that characterised Francoism. The fact that his body has never been found, meanwhile, has come to represent the buried and unsettled legacy of the civil war and subsequent dictatorship. For many, he's a disappeared person and represents all the dead of the Spanish civil war and all the horror of the war and the dictatorship.

Not sure that was quite reaction Bob expected from his opening slide.

Bob continued, "We've been fighting the global war on terror for many decades. We all understand here that advanced surveillance methods and other techniques developed in the counterinsurgency wars are migrating from Baghdad, Falluja and Kandahar to hometowns and urban neighbourhoods."

"And don't ever claim that nobody told you this could happen."

"Brant think of the counterinsurgency wars abroad as so many living laboratories for the undermining of a democratic society at home, a process that has been going on for a long, long time."

Rachel Jackson interrupted, "Steady, Mr Ranzino. Don't you think that 'living laboratories' is an excessively emotive term to describe what you are doing?'

Bob answered, "Surely, Rachel, you know it is not my term? The military has been using it for a long time to describe the counterinsurgency wars. Sure, we've counterintelligence innovations like centralised data, covert penetration and disinformation developed during the Army's first protracted pacification campaign in a foreign land—the Philippines from 1898 to 1913. But these techniques were repatriated to the United States during World War I, becoming the blueprint for internal security apparatus that persisted for the next half century."

"That's strong stuff, Bob. Are you sure that the Pentagon would agree with your assertions?" asked Dan Anderson.

Kyle interrupted in our listening area, "Wow, Bob is dragging some dark stuff into the discussion. That was a bloody war where the Filipinos brought knives to a gun fight."

Bob continued, "Think about it. With the US military plunged into four simultaneous counterinsurgency campaigns, large and small–in Somalia, Iraq, Afghanistan and the Philippines–transforming a vast swathe of the planet into an ad hoc 'counterterrorism' laboratory."

"You're using that word again - laboratory. It makes the warfare sound like a giant experiment," said Rachel.

Bob answered, "Well, Rachel, it was. Look at the results? Cutting-edge high-tech security and counter-terror techniques that are now slowly migrating homeward. Brant has been asked operate on behalf of so many campaigns both on home soil and in the managed facilities of other countries."

"Boots on the ground in foreign climes?" asked Dan.

Kyle laughed, "Boots on the ground, Bigsy. I think I win the military Lotto."

Bigsy laughed, "But will he use the words surveillance or data-mining?"

"Boots on the ground both abroad and at home," replied Bob, "Face it. Every American knows that we fight elsewhere to defend democracy at home. Yet the crusade for democracy abroad, largely unsuccessful in its own right, has proven remarkably effective in building a technological template that could be just a few tweaks away from creating a domestic surveillance state–with omnipresent cameras, deep data mining, nano-second biometric identification, and drone aircraft patrolling 'the homeland.'

I winced when he used the word crusade. I wondered if the two news anchors would pick it up. Bob was manipulating it to sound ultra-right and totalitarian.

Dan spoke, "But wasn't that whole democratisation push really about oil and energy? Diplomatic visits to whichever states could give the US favourable terms for oil?"

And Rachel, "The sad, mildly ironic reality of the US approach to democracy promotion is that critics say it may represent the worst of both worlds: it has soured people all around the globe, and many in the United States as well, on the very legitimacy and value of US democracy promotion."

Bob spoke again, "But Rachel. Let me continue: The ongoing support Brant Industry brings has helped a massive expansion of domestic surveillance by the FBI and the National Security Agency (NSA) whose combined data-mining systems have already swept up several billion private documents from US citizens into classified data banks. Now we can understand where this dissent stems from and take steps towards correction."

Rachel spoke, "But that sounds almost dictatorial?"

Rachel spoke to Amy, "Is he always like this with women? Only he seems to be talking across Rachel and ignoring her questions."

Amy answered, "I've always assumed it was simply me he didn't get on with, now I can see it is more his general demeanour."

Bob replied, "No, it is populist, Rachel. Populist. What the electorate want. The Second Amendment writ large, but safely. Let me give an example: Abroad, after years of failing counterinsurgency efforts in the Middle East, the Pentagon began applying biometrics – the science of identification via facial shape, fingerprints and retinal or iris patterns – to the pacification of Iraqi cities, as well as the use of electronic intercepts for instant intelligence and the split-second application of satellite imagery to aid an assassination campaign by drone aircraft that reaches from Africa to South Asia."

Rachel asked, "But doesn't that sound like an Orwellian Big Brother scenario?"

Bob replied, "Certainly not. It means, in the panicky aftermath of a future terrorist attack, Washington can quickly fuse existing foreign and domestic surveillance techniques, as well as others now being developed on distant battlefields, to create instant digital surveillance. What is needed is an ability to scan large areas - just what Brant is now developing in its HCCI Artificial Intelligence product line."

"I'm sorry but what Bob describes sounds just like Big Brother to me," said Juliette, looking annoyed.

Dan asked, "HCCI?"

Bob replied, "Oh yes, I'm sorry - I guess this audience knows HCCI, but it isn't really a layman's term. Human to Computer, to Computer to Human Interface. The way to speed the links between two Humans, boosted by the intermediate application of Artificial Intelligence. "

Rachel asked, "So does that mean that the computer can override the human? In either direction?"

Bob answered, "To a point, but the conventional instincts of a human will still override any strange or illogical commands from the computer. Like in the case of a computer command to go into a fire, the human instinct would win."

"Finally, a civil answer to one of Rachel's questions!" said Amy.

Bob continued, "This new leveraged human system was first mooted in the smoking aftermath of Operation Phantom Fury, a brutal, nine-day battle that US Marines fought to recapture the insurgent-controlled city of Falluja."

"Bombing, artillery, and mortars destroyed at least half of that city's buildings and sent most of its 250,000 residents fleeing into the surrounding countryside. Marines then forced returning residents to provide their fingerprints and iris scans. Once inside the city's blast-wall maze, residents had to wear identification tags for compulsory checks to catch infiltrating insurgents."

Rachel asked, "Yes, but the residents had to wait endless hours under a desert sun at checkpoints for this to be achieved. Some could say it disregarded the population's feelings entirely."

Bob spoke, "Yes, Rachel, you are reinforcing my point. Now, with Brant's new technology, we can sweep across an area, collecting biometrics, face and body profiles. We can classify people in quantity and then using peacekeeper forces like those supplied by Brant, the US can check Iraqi identities by satellite link to a biometric database."

"He's talked across her - again!" said Amy, "I wonder whether there are many women listening to this?"

Dan questioned this statement by Bob, "Isn't it an excuse to use mercenary power to manage a war-torn state, in the name of the US?"

Bob spoke, "Not at all. This initiative was extended with Joint Expeditionary Forensic Facilities, linked by satellite to a biometric database in West Virginia. A war fighter needs to know one of three things, 'Do I let him go? Keep him? Or curtail any further action?'"

Rachel asked, "That last item, it sounds final?"

Bob continued, "Rachel let me explain."

"Mansplain more like!" said Amy, "Godverdomme!"

Bob continued, "With Brant's help, American intelligence agencies launched a Special Action Program using the most highly classified techniques and information in the US government to locate, target and neutralise key individuals in extremist groups such as al Qaeda, the Sunni insurgency and renegade Shia militias. Without Artificial Intelligence and Image Recognition - we say AI-

IR - this would not have been possible. Today, in the demonstration, we can show you some of the latest technology we are working on. There's a couple of big ideas."

Dan said, "Big Ideas, against this backdrop. You'll have to explain."

I could see that the presentation had reached its pivot. It had run through FUD - Fear, Uncertainty and Doubt and was moving into the Case For Change - The S Curve Jump.

It intrigued me to see how Bob would pull this off.

Bob spoke, "Mark Twain nailed it when he said 'You know, I'm all for progress. It's change I object to.'"

Now I've been a consultant too, so I could sense the moves. Bob was going to create a simple, compelling case for change.

Kyle, sitting with us, said, "Ha, there's a turn up. Twain was famously against that Filipino War that Bob used in his case earlier. Twain said the war betrayed the ideals of American democracy by not allowing the Filipino people to choose their own destiny."

On stage, Rachel asked, "Bob, I can see that you are fired up on this thing, but won't there be many doubters?"

Bob spoke, "Rachel, this isn't for the complacent. Look around and there will be people prepared to sit on their hands. Maybe you are sitting on yours, Rachel? This is for those of us who are striving to reach further. We know that there is a window of opportunity open today, but that might be about to close. We can move to automate the scanning, ahead of any hue and cry about civil liberties. Those that don't will crash and burn. They will never have the knowledge of their citizens. Like fields of wheat, we can this way avoid wheat-blast, the contamination caused by a few bad grains, nibbled by insects, or worse, by insidious viruses."

Rachel kept her composure and asked again, "You are saying that this is the time to act, but using what?"

Bob continued, "Rachel - that's right. We need to manage the outcomes we discover. That's where we link the CC computers together at the heart of HCCI. We'll be able to administer that crop better if we can understand and translate it."

Dan asked, "Bob, by crop - I assume you are using a metaphor?"

Bob spoke, "Dan, Correct - I am using a metaphor. For this to work, those that are empowered must hit every channel we can think of. All channels. That means talking at town halls, team meetings and as we walk around. Write a leader's blog or send an email. Use powerful visuals and language. Be passionate! People can feel it and it is contagious."

Dan "So it's like putting on a gorilla suit with a label on your back?"

Kyle, Bigsy and I looked at one another - it was subtle,

but Dan was having a jibe at Bob - probably about his treatment of Rachel.

Hermann laughed to Rolf, " A gorilla suit - Ha!"

I realised that the language and humour setting had to be dialled to 'Mr Bean' for it to work with my two German friends.

Bob laughed, "You could do that, Dan. I say keep it simple, keep it honest and straight. The spin doctors will kill the message by covering up hard truths with pretty words. People smell that aroma a mile away and if they don't, they will hate you when they find out the truth and whammo! FAIL! You have just created an Everest to climb. Trust is essential to killing the Fear, Uncertainty and Doubt.

Rachel said, "This is all very well, but they say it takes a village to raise a child. Systemic change requires pressure to be applied at many points? How will this achieve it?"

Bob commented, "The hero in the organisation can sometimes make a difference in the system. But it is not sustainable and when the hero burns out or leaves, the driver of change burns out and leaves with them."

Rachel asked, "You haven't answered my question, you are still talking about a single hero?"

Bob answered, "No, Rachel, I'm not. Meaningful organisational change requires a guiding coalition of champions. Enlist a core group of champions for change. It's a lot easier to drive with an army of advocates than it is as a singular voice."

Hermann spoke quietly to Rolf, "I can understand what 'buzzword compliant' means today!" they both laughed.

Bob leaned toward Rachel, "Rachel, you know what? The teams everyone brought along to these sessions could be the best ones to start something. Each own organisation will have selected you, to be the best or the most knowledgeable about the topics we are here to discuss. Right Now."

Bob briefly stood, for his next remark, "Leaders, you'll need to invite some of your people to be a part of a meaningful opportunity to create change. Share the vision and build the belief of others to magnify its power. Connect with those affected one-on-one to make it relevant, understand their concerns and ensure everyone is clear. The lesson is clear. If you develop the new S-Curve and develop the new program for change and simply hand it to your people, saying, 'Here, implement this,' it won't work."

Dan asked, "Bob, the way you paint this, some could see it as dystopian, terrifying even. How would you handle this resistance?"

Bob replied, "There will be barriers, there will be resistance. There will be pain and suffering. There will be eye-rollers, detractors, underminers and fear sowers. The FUD monster will run amok. Take this as a sign, a sign that you are on the right path. A sign that the efforts of change are working to stir up the resistance. This is when you need to press harder."

Dan asked, "But are you not simply forcing through your own worldview?"

Bob answered, "I am forcing through the view that is in the organisation's best interest. The need to adapt. To question the status quo – kill off 'that's the way things have always been done around here'. Recognise, reinforce, celebrate, and reward. Create positive pressure that shines a light on the new behaviours and calls out old ways of doing things."

Dan, "But some may simply not see it your way?"

Bob, "Make no mistake, you'll need to deal with unyielding resistors, the 'smiling assassins' will need to be nipped in the bud. You can't forget: Everyone is watching what happens; they want to see who will win— you or the resistor. Your actions have great symbolic significance, and how you deal with resistors can determine the success of the entire change effort."

Rachel asked, "So what are you going to do with these resistors?"

Bob answered, "I would have thought it was obvious - get rid of them. But How? These people each have their own fan club of ego boosters. We must give them tough love. Give them a chance to respond. But remember it is tough love. People that resist the new aren't wrong, they might just be better off working somewhere that is a better fit with their worldview. Unless they adapt, move them out."

Hermann and Rolf were struggling with 'Tough Love'; their translation seemed to mean something different. Rolf muttered, "Nja, es ist besser in Englisch"

I knew the next thing would be about celebrating early victories and sure enough, Bob's train kept on coming.

Bob continued, "Jumping an S-Curve can take a while. Large-scale change can be a long, formidable undertaking, so it is important to create short-term wins. Several early victories, even if they are small, create self-confidence and the belief that bigger successes are possible. This belief builds a psychological momentum that sustains the effort needed for large-scale, long-term change. That's what you'll need to show - Wins - and a win can be anything – big or small – that helps you move toward your opportunity. They take many forms - courageous actions, processes improved, new behaviours demonstrated. These wins show the change is working."

'Ooh Jasmine,' I thought, 'You've copied too much of the textbook.'

"And who will stand up for the first of these small wins?" asked Dan.

Bob replied, "Dan, you know what? You'll need to be out front. Show a genuine enthusiasm that generates energy - enthusiasm that is based on your belief in what you are doing. Display a clear strategic focus and show your people you have confidence that they can execute the plans. This confidence and calmness is contagious. It is the opposite of FUD and will slay it."

Bob stood briefly and looked into the auditorium, "Go ahead and shine. You know you have it in you. The success will come, and Mark Twain will be on board beside you smiling."

There was a light smattering of applause after the last statements, although I realised that Bob, or probably Jasmine had probably repurposed a PowerPoint slide deck for that entire section of Bob's talk.

I assumed that was the end of the second act of Bob's talk and it would be sell, sell, sell from here on in. After all, he'd told us what he was going to tell us, then he'd told us, and now he was going to tell us what he'd told us.

Bob smiled appreciatively at the ripple of applause.

"I can see there's a few Leaders in the house today! Now let's talk some more about the things we've been doing at Brant."

'Here it comes,' I thought.

Bob continued, "A crucial technological development in Washington's secret war has been the autonomous drone, or unmanned aerial vehicle, whose speedy development has been another by-product of Washington's global counterterrorism laboratory. No need to ask the question again Rachel!"

A few people around the audience laughed. I found the totalitarian, no compromise positioning rather unfortunate and now he was browbeating his woman interviewer, things were entering further troubling territory.

Bob continued, "The Pentagon proposed an expenditure of $1.2 billion for a fleet of fifty light aircraft loaded with advanced electronics to loiter over battlefields in Afghanistan and Iraq, bringing full motion video and electronic eavesdropping to the troops. Of course, that was ahead of Brant's developments in this field, but nowadays they could have full Facial Recognition with Artificial Intelligence - FR-AI. This isn't just for offshore but could become a major benefit for domestic policing also. We'll run a demonstration of this today, based upon the fully robust system we use for crop inspection."

"Now, it doesn't take much imagination to see how the HCCI link from this intelligence gathering could also be applied to both an information gathering autonomous drone but also to something like a Predator cruise missile."

I could see Bob was getting it ready to be a sales proposition.

"Then we have joined-up missions, with Humans moderating the interactions between a target acquiring drone, and a sophisticated aerial mission platform like a Predator. Full HCCH managed missions. Let's call that Idea One."

Kerching.

Bob continued, "Then we get to CITVIG. While those running US combat operations overseas were experimenting with intercepts, satellites, drones and biometrics, inside Washington the civil servants of internal security at the FBI and the NSA initially began expanding domestic surveillance through thoroughly conventional data sweeps, legal and extra-legal."

Dan asked, "Bob - you'll have to slow down, what's CITVIG?"

"Oh! Citizen Vigilance," Bob continued, "An example: The US Justice Department launched Operation TIPS (Terrorism Information and Prevention System), with plans for millions of American truckers, letter carriers, train conductors, ship captains, utility employees and others to aid the government by spying on their fellow Americans. Of course, without the LCM - Leadership Change Management we've been discussing here, the program had to be formally buried. Notice I say 'formally', because we can reinvigorate it at a stroke with HCCI. Computer assisted inspection of relevant data streams with the results automatically uploaded to a CITVIG database in the cloud. A second HCCI can sift and action the intel received far more efficiently than plodding civil servants ever could. Let's call that Idea Two."

Kerching again, I thought. But Bob seemed to be playing with alphabet soup to the American audience rather than the vastly wider global one. And he was probably treading on a few toes too.

"And if we peep inside the Pentagon, the original Admiral Poindexter Total Information Awareness program - that's TIA - was to contain detailed electronic dossiers on millions of Americans.

"But wasn't Poindexter discredited?" asked Rachel, "Something to do with the Iran-Contra affair?"

"I think he was later re-instated to DARPA, actually," answered Bob, "But we are straying off topic,"

Another side swipe at Rachel.

Bob continued, "When news leaked about the secret Pentagon TIA office with its pyramid, all-seeing eye and globe logo, Congress banned the program, and the admiral resigned. But the key data extraction technology, the Information Awareness Prototype System, migrated quietly to the NSA. But no-one had the AI to process the findings. Until now. There's an Idea Three."

Kerching - Bob was really going for it.

Bob, "And with even rudimentary systems, the CIA, FBI and NSA turned to monitoring citizens electronically without the need for human tipsters, rendering the administration's grudging retreats from conventional surveillance at best an ambiguous political victory for civil liberties advocates.

Dan asked, "But do you take the civil liberties stance seriously in Brant?"

Bob replied, "We at Brant can't condone lawbreaking. Let's face it, the NSA were given secret, some might say illegal, orders to monitor private communications through the nation's telephone companies and its private financial transactions through SWIFT, an international bank clearinghouse. After the New York Times exposed these wiretaps, Congress quickly capitulated, first legalising this executive program and then granting cooperating phone companies immunity from civil suits. So, I guess you could say it was illegal for a while, but then legislation caught up with the modernity of thinking. And a part of it does still rankle. Such intelligence excess was intentional. The term for the activity was, systematic "over-collection" of electronic communications among American citizens."

I was impressed at how Bob had woven his way through that little question, still delivering his payload in his answer, but appearing to be on the right side of the law.

Bob continued, "But time moves on and attitudes change. Now thanks to a top-secret NSA data base called 'Pinwale' analysts routinely scan countless millions of domestic electronic communications without much regard for whether they came from foreign or domestic sources. The wiretap network is called DNI - The Digital Network Intelligence. There's a challenge though. The data is so voluminous, it cannot be held for long. But maybe add in an HCCH interface and the determination of which pieces should stay and which pieces could go, becomes vastly simpler. Let's call that Idea Four."

'Bob,' I thought, 'Stop. You have made your point.' But oh no, he just kept on.

Bob added, "I could go on and talk about the FBI with its Investigative Data Warehouse as a centralised repository for counterterrorism. Within two years, it contained 659 million individual records. This digital archive of intelligence, social security files, drivers' licenses and records of private finances could be accessed by 13,000 bureau agents and analysts making a million queries monthly. By 2009, when digital rights advocates sued for full disclosure, the database had already grown to over a billion documents."

Bob added, "And did this sacrifice of civil liberties make the United States a safer place? No - because it was reactive. You'd find someone after something has happened, but never before. I know what you are all thinking...HCCI Idea Five? Well maybe, just maybe."

Dan stood, "Hey everyone, we've reached time. I think we'd like to thank Bob for this keynote, which has certainly fired up the whole Expo. He talks of many controversial topics and provides a mind-expanding range of solutions."

Then Rachel, "Bob has certainly raised a few civil liberties issues during his talk, the very idea that offshore battlefields could be experimental testing grounds for citizen management systems at home raises various both moral and ethical challenges, I hope we get a chance to debate this some more with Bob and other senior members of Brant."

Bob stood up again, "This has been a whirlwind tour. I've presented you with some issues, a fearless project change process and then a few illustrations of ways to help you all with your individual initiatives. You've been a great audience and I can tell you've enjoyed the presentation. I hope to see some of you drop by the Brant Exposition

Stand. We've a Breakout room there too and some of you might want to drop around, by appointment, to our executive briefing suite in the Hotel Rey Juan Carlos I - You can get your invitations via our display area. And I'd also like to thank Dan Anderson and Rachel Jackson, two spirited TV presenters who have certainly given me a run for my money! I've been Bob Ranzino and don't forget - 'The Future is Brant!'

Bob stood again, stepped forward and bowed with both arms outstretched. A true professional. He got the applause he was expecting.

The Future is Brant

After Bob's presentation, the exposition attendees made their way out. They had a choice of routes, but many siloed out into the Expo area and past our stand. It was incredible just how many people were stopping for a look. Either an immediate look, or to book a separate quieter session, either in the breakout room or back at the Executive Briefing Suite.

We had the little scanners so we could work out which people were influential, even if it was not clear from their chosen, sometimes military, uniforms.

Chuck, Jake, Christina, and Clare returned.

"That Bob," said Christina to Amy, "He's a bit of a misogynist, isn't he? The way he shut down Rachel every time she tried to answer a question!"

"Yes, and theoretically she could have been a political supporter of Brant, if it hadn't been for that very direct action by Bob."

"But who gave him the spiel?" asked Clare, "I mean, some of that was Straight Outta Handy. - Or even earlier." Clare referred to Charles Handy's treatise on S-Curves.

"We should get some tee-shirts made," joked Jake, "In Bigsy's style 'iJumped' or something like that?"

"Or 'The future is Brant'," laughed Hermann. In German it means "The Future is blazing! We could include some artistic flames!"

Rolf and Hermann were taken aback by the number of requests for demonstrations of the HCCI, using one of the experiments in which I featured. They set up the breakout zone as a mini theatre and could seat 30 people at a time. I watched as they improvised a queuing system like the one at a deli counter in a supermarket.

I could also see that a couple of those Chinese policemen were loitering around the stand. They both had the XXL sized hats and electronic dark glasses, which I assumed were being used for facial recognition.

Kyle nudged me, "Remember that CaptureIt thing we built back in the casino days?" He was referring to a small device which could pick up digital signals and copy them. We'd used it to experiment before we'd built the first crypto engine.

I realised what Kyle was doing. He had set up CaptureIt to reclaim the recordings being made by the Chinese.

After about thirty minutes, the Auditorium had emptied, and things had become quieter. Rolf and Hermann were in mid demonstration and Amy with Juliette was handling the stream of enquiries from Bob's presentation.

Bob Ranzino was at a Press Conference where the great and the good could ask him more questions from his presentation. The Expo had its own 4-pager which it handed to delegates on their way into the sessions, and I assumed that next day's version would be un-judgementally full of what Bob had told them.

Then, Juliette unhooked herself from the enquiry desk and walked over.

Juliette tells me things

"Hey," Juliette said, "It's been crazy here, hasn't it?"

I nodded, and she said "I've found out some things though..."

"Bob's wife, Mary, is setting up a session for Brant in Beijing. There's another conference like this one, at the Bei Jing Zhan Lan Guan - That's the New China International Exhibition Center, which is quite close to where Mary worked, when she was on loan at the Peking University."

"Why would they want a repeat of this?" I asked, confused.

"Think about it. The Chinese internal market is vast. There are 2.1 million soldiers in the Chinese Army. Their spending on military is second only to the USA, at $261 billion US dollars. America is still the largest at $732 billion, but when you look at Russia $65 billion or any European Country (less than $50 billion) then you start to see why Brant sees the Chinese market as so attractive."

"Is that why Bob's speech was so -er- biased?" I asked.

"You mean totalitarian?" smiled Juliette, "As a matter of fact, I asked Jasmine Summers and she admitted that she'd written it. It wasn't so much a political diatribe, as a naive cut-and-paste from a couple of oven-ready presentations."

"Well, it will have pushed the right buttons among the Chinese," I said, "The Chinese approach to conflict goes right back to Sun Tzu. It's ingrained. They want the totality of foreknowledge to operate. That's why their information gathering is so important. Whether it is of the home citizenry or a foreign force, they want to know as much as possible about them, so that they can enact a strategy."

"That'll be why they have asked for a VIP Briefing here, then, and the follow up of the visit to Beijing. Mary Ranzino has already asked Qiu Zhang to coordinate the sessions with the Chinese."

I made a mental note to check with Grace for any further information about Qiu Zhang.

"Oh yes, and Chuck suggested we all get together in a bar tonight to compare notes," Juliette added.

Chuck walked over, "Beach Life, " he said, "It's a shack just along from the W Hotel."

It reassured me that even Chuck was planning drinks rather than security shakedowns.

Replacement of reality

Rick Deckard: *Is this a real snake?*

Zhora: *Of course, it's not real. You think I would be working in a place like this if I could afford a real snake?*

Bladerunner

SinoTech

The afternoon's keynote was from SinoTech, the Chinese technology manufacturing company. They would be displaying several of their leading-edge smart systems, from a modest domestic cleaner through to their latest commercially available security systems.

As a group, Brant decided to visit this session, leaving Juliette and Jake to mind the stand.

"It's okay," said Jake, "I got the drift from this morning's session. I don't think I can take any more authoritarianism."

"I'm good too," said Juliette, "I can watch the session on one of the monitors. It will be nice to be away from the crowds for a while."

The rest of us trooped into the Auditorium and took a block of two rows, to the left hand side of the main auditorium. I was surprised to see how packed the session was, and I did also notice several Chinese-looking faces around our area.

I was in the row with Kyle and Bigsy, in front of us was Rolf, Hermann and Amy and to our right sat Chuck,

Christina and Grace. Unusually, I noticed Christina was wearing a baseball cap, with a Brant logo.

The session began, and SinoTech began to introduce a stream of devices. They had chosen a presenter - Huang Li - who spoke good English, although her voice was in an upper register and was really quite piercing.

She started off by showing a domestic robot vacuum cleaner. It was remarkably similar to a Roomba but was around one third of the price. It had also been bought by many families in the USA, thanks to a discounted programme on Amazon.

Huang Li referred to Bill Gates and his Microsoft mission during this session, "A computer on every desk and in every home."

Huang Li said that when Gates had first described the mission it was almost unimaginable but now it was passé. It was so time-limited that Microsoft CEO Satya Nadella, who started at the company as a programmer in 1992 and rose to the top job in 2014, thought Gates' famous mission had a big flaw. Nadella redrafted the mission to being "to empower every person and every organisation on the planet to achieve more."

I thought 'blech' at this Nadella one, although I honestly remembered the old Gates' one. Bill Gates probably doesn't care, he is still making $380 per second, or $12 billion per year.

And now, Huang Li proposed the SinoTech megamix, "Multiple intelligent devices in every office and home, to help people achieve more."

I could see where this was going, cloning a Roomba and now copying a corporate mantra.

Kyle, muttered something, "I see, sensors in every home? Smart homes filled with electronic surveillance?"

Of course, we were being cynical, but I guess Kyle had a point. The Americans were trying to stop the influx of Chinese 5G systems onto their national infrastructure. They had ignored all the other devices that consumers would buy anyway, like phones, doorbells, intelligent switches, vacuums and so on.

The next thing that Huang Li brought out looked a little like a parking meter, except it moved around. It was a store aisle monitor. Intelligent in the sense that it could look for spillages. Dumb in that it could not 'Roomba' them up. I wondered if this was a clever way to run surveillance inside supermarkets. Add in some facial recognition and maybe some bluetooth detection?

Kyle, Tyler and Bigsy were chattering, "We'll have to do a teardown of one of those!"

Huang Li showed various skins that could be applied over the contraption to make it look cuddlier. Some US Stores were already selling plush toy replicas of the device. It reminded me of that pesky broomstick in Fantasia.

Then she brought in a stockier version of the same device, designed for use in malls. The Brits were all thinking the same thing,

"Oh My God, It's a Dalek."

Well, it wasn't of course, it was shorter, and more bullet shaped than the Doctor Who enemy known as a Dalek, maybe based more on that silver and blue dustbin robot in Star Wars?

This was overtly more of a surveillance system, for use in malls and had a speaker array which could call out to anyone creating trouble. It had 360-degree cameras and could patrol autonomously.

"I bet it can't go upstairs!" said Bigsy.

Then on to their smaller cart type devices. These could move fast and had cameras on them as well as undefined sensors. Suitable for rapid deployment, they could also be used on streets. It reminded me of one of those expensive electric car toys for children, but with extra cameras. It was explained it could also provide parking tickets.

I could sense that Chuck was sizing them up to work out how much force would be needed to flip them over or to really disable them.

Then we were into the realms of food and package delivery, although I wondered how many pizzas would get stolen from them if they were a purely passive device?

Huang Li switched on the video for the next devices. One

was a flying unit, essentially an inexpensive surveillance drone. Then a dove-shaped drone, which was supposed to provide the same results but with less menace. It turned out that the doves were already in use in Xinjiang. And then a couple of large jeep-styled vehicles, which were fully wired, and which could patrol perimeters like borders or walls.

Huang Li didn't have the same summarisation skills as Bob though and the end of the session ended flat, with many in the audience feeling ever so slightly threatened by what they had just witnessed.

Nonetheless, there was hearty applause as Huang Li left the stage.

Kyle spoke to me, "You know, we should take up my friend John Yarvay's offer to present us a mini pitch on Chinese security."

I could see Grace had picked up on Kyle's comment and was nodding, as was Chuck.

"Good idea," said Grace, "We need to cut through all of this."

Xuan Yahui

We were using the Brant VIP Briefing Suite for John Yarvay's presentation to all of us. Bigsy had theatrically been around it with his 'lollipop' scanner, to check for any bugs.

We had Amy, Juliette, Rolf, Hermann and me representing Brant and then Chuck, Christina, Bigsy, Clare, Jake, Kyle and Tyler representing the security services. Amanda had to go to another meeting somewhere, but it was still quite a crowd. We'd decided it was best not to involve the Brant top brass in this session, which would, after all, be talking about them.

So neither Bob nor Mary Ranzino, also no Jasmine Summers.

Kyle introduced John (Xuan Yahui) to Grace, and she attempted to say a few words in Mandarin back to him.

"Fēicháng gǎnxiè nín wèi wǒmen suǒyǒu rén zhuànxiě de dàliàng lùnwén."

(Thank you very much for writing many papers for all of us.)

John looked confused," My pleasure, and thank you for attempting to thank me in Chinese." He grinned but I realised he was not going to say anything about the odd translation.

"Kyle and I go back a long way. Not as far as Matt and Kyle, but almost. We've both helped each other out of a few scrapes in the past. Kyle is exceptionally good at security you know."

"I'll give you the money later, " replied Kyle. Hermann laughed.

Kyle introduced John and said the two of them would sometimes sit together in bars and make up new security jokes:

"Some examples:

Q: What's a hacker's favourite season?
A: The phishing season.
Q: Why did the programmer go to rehab?
A: He was addicted to coding.
Q: What's the best way to catch a runaway robot?
A: Use a botnet.
Q: Why did the Programmer leave the campsite early?
A: There were too many bugs."

There was some concentration around the tables as people tried to work to what the jokes meant, but no real laughter.

"You're a tough crowd, but now, please let me present John Yarvay!"

John began, "Thank you Kyle and, well, at least we've narrowed the Brant security risks down to two groups," he began,

"1) Everyone who works here and
2) Everyone who doesn't work here."

Hermann laughed at this one.

"It's great to present to you all, and this deck has stood the test of time, as a general primer of how best to consider the security actions of the Chinese. I should state up front that my stance on this is apolitical. I should declare myself as originally from China but having spent twenty years living in the United Kingdom and then travelling the globe working as a security advisor."

"As an example, how many of you noticed that the cute parking meter in the SinoTech presentation was running a comprehensive facial scan of everyone present? Or that there are several Chinese policemen over in the hall and in this hotel's lobby who are running data acquisition?"

John started the PowerPoint and showed the 'parking meter' shaped device, with its camera on top. Then he showed a picture of two of the large-hatted police, both pressing the sunglasses on their heads to capture pictures.

I suddenly realised why Christina had been wearing that

baseball cap and sunglasses inside the SinoTech session. Not a hangover, simply a security precaution.

John flicked onto slide 3- a blue logo - stating C.A.S.:

"Here we are - Northwest of Beijing's Forbidden City, outside the Third Ring Road, the Chinese Academy of Sciences has spent seven decades - seventy years - building a campus of national laboratories. Near its centre is the Institute of Automation, a sleek silvery-blue building surrounded by camera-studded poles. The institute is a basic research facility.

Its computer scientists enquire into artificial intelligence's fundamental mysteries. Their more practical innovations—iris recognition, cloud-based speech synthesis—are spun off to Chinese tech giants, AI start-ups, and, in some cases, the People's Liberation Army.

"I visited the institute on a wet morning last year. They wanted to interview me. I think they wanted to recruit me to work for them in the west. I was there early and China's best and brightest were still arriving for work, dressed casually in shorts or yoga pants. In my pocket, I had a Mars Bar shaped burner phone; in my backpack, a computer wiped free of data—standard precautions for Western researchers and journalists in China."

Grace noted, "Yes, to visit China on sensitive business is to risk being barraged with cyberattacks and malware. As an example, we know of Belgian officials on a trade mission noticing that their mobile data were being intercepted by pop-up antennae outside their Beijing hotel."

John smiled, "There's plenty of those antennae around nowadays, actually. They are now small enough to fit onto cars and look almost like regular aerials."

"Here:" He flicked to a picture of an unrecognisable and bland looking Chinese car, which seemed to have four aerials.

John continued, "I had to be processed by the institute's security and then I was told to wait in a lobby also monitored by cameras. On its walls were posters of China's most consequential postwar leaders. Mao Zedong loomed large in his characteristic four-pocket suit. He looked serene, as though satisfied with having freed China from the Western yoke."

Clare interrupted, "It makes me shudder. Although Mao has been credited with transforming China from a semi-colony to a powerful sovereign state, he ruled an autocratic and totalitarian regime responsible for mass repression, and destruction of religious and cultural artefacts and sites. His state was additionally responsible for vast numbers of deaths with estimates ranging from 40 to 80 million victims through starvation, persecution, prison labour and mass executions."

Chuck mused, "The power of propaganda to paint over the crime?"

John continued, "Next to him was a fuzzy black-and-white shot of Deng Xiaoping visiting the institute in his later years, after his economic reforms had set China on a course to reclaim its traditional global role as a great power."

This time it was Grace who interrupted, "Deng's policies gradually led China away from a planned economy and Maoist ideologies, opened it up to foreign investment and technology, and introduced its vast labour force to the global market. Such actions are credited with developing China into one of the fastest-growing economies in the world for several generations and raising the standard of living of a quarter of humanity.

Grace continued, "He was eventually characterised as the "architect" of a new brand of thinking combining socialist ideology with free enterprise, dubbed 'socialism with Chinese characteristics'. This ideology now known as Deng Xiaoping Theory was ultimately incorporated into the party's constitution in 1997 and remains the guiding principle of Chinese policy to this day."

John again, "But by far the lobby's most prominent poster depicted Xi Jinping in a crisp black suit. China's president and the general secretary of its Communist Party takes a keen interest in the Institute.

John flipped to a new slide which showed a Future of Life Institute logo with a Red Chinese flag underneath which were the words: New Generation Artificial Intelligence Development Plan (新一代人工智能发展规划).

John continued, "The institute's work is part of a grand Artificial Intelligence strategy that Xi has laid out in a series of speeches. Xi has said that he wants China to be competitive with the world's AI leaders, a benchmark the country has arguably already reached. And he wants China to achieve AI supremacy."

Juliette nodded, "Yes, we've seen some of the papers from that Institute. There's one guy who publishes a blog like a newsletter and he'll translate some of their work and the findings that they have achieved. The tone is always very upbeat."

Grace interrupted, "I've seen that too. But Xi's pronouncements on AI have a sinister edge. Artificial intelligence has applications in nearly every human domain, from the instant translation of spoken language to early viral-outbreak detection. But Xi also wants to use AI's awesome analytical powers to push China to the cutting edge of surveillance. He wants to build an all-seeing digital system of social control, patrolled by algorithms that identify potential dissenters in real time."

"Is that like the Precogs in Minority Report?" asked Jake, "You know when they predict the crimes in advance? - Philip K. Dick style?"

Kyle answered, "Yes but Precrime units are still fanciful concepts. There's a big jump between going to the 'murder crime' book group and becoming a murderer. Like watching Scandi-noir TV shows doesn't make someone a criminal in waiting."

John continued, "There's some subtleties to China's behaviour though. China's government has a history of using major historical events to introduce and embed surveillance measures. In the run-up to the Olympics in Beijing, Chinese security services achieved a new level of control over the country's internet. During China's coronavirus outbreak, Xi's government leaned hard on private companies in possession of sensitive personal data. Any emergency data-sharing arrangements made behind closed doors during the pandemic could become permanent."

Kyle said, "And it is well-known that China already has hundreds of millions of surveillance cameras in place. Xi's government hopes to soon achieve full video coverage of key public areas. Much of the footage collected by China's cameras is parsed by algorithms for security threats of one kind or another."

"That's like the Precrime concept," said Jake.

John continued, "In the near future, every person who enters a public space could be identified, almost instantly, by AI matching them to an ocean of personal data, including their every text communication"

Juliette again, "Actually they were doing something like that on the SinoTech stand in this Expo. When Amy and I walked around I assumed they were using our Exhibitor badges to tag us all onto their big screen behind, with our names and companies."

Amy nodded, "Yes, they had both of us correctly identified. From what I'm hearing now, I guess we have been pulled into some giant Chinese database!"

John continued, "In time, algorithms will be able to string

together data points from a broad range of sources—travel records, friends and associates, reading habits, purchases—to predict political resistance before it happens. China's government could soon achieve an unprecedented political stranglehold on more than 1 billion people."

"I'm not convinced, " said Clare, "The way Facebook treats me suggests it hasn't really got much of a clue about my profile. Last week it was sending me adverts for cowboy boots. The week before it was spaghetti drying stands."

Jake laughed, "Maybe it thought you liked spaghetti westerns?"

Everyone laughed.

Bigsy spoke, "Yes, Clare, but you are savvy. You know how to throw the algorithms off the trail. Look up something ugly or bad and watch Facebook attempt to fold it into your good profile."

Jake and Chuck laughed this time, but I sensed it whistled past my European colleagues.

John replied, "But it can be simple ideas that win through. Early in the coronavirus outbreak, China's citizens were subjected to a form of risk scoring. An algorithm assigned people a colour code—green, yellow, or red—that determined their ability to take transit or enter buildings in China's megacities. In a sophisticated digital system of social control, codes like these could be used to score a person's perceived political pliancy as well. Traffic lights for their levels of access to capabilities."

"What? So it would score me based on what it thought I believed?" asked Clare, looking annoyed.

John said, "Already there's a crude version of such a system in operation in China's north-western territory of Xinjiang, where more than one-million Muslim Uighurs have been imprisoned, the largest internment of an ethnic-religious minority since the fall of the Third Reich. Once Xi perfects this system in Xinjiang, no technological limitations will prevent him from extending AI surveillance across China. He could also export it beyond the country's borders, entrenching the power of a whole generation of autocrats. You could call it an experiment like Bob Ranzino described in his rather worrying keynote speech."

Rolf asked, "You mentioned the megacities? Do they exist already?"

John put up another slide. It was a cityscape, "Take a look at some of China 2.0 - the birth of the Megacity. Shanghai, Beijing. Jing-Jin-Ji. Tianjin. All emerging as Mega Cities. Shanghai population is around 30 million. China has embarked on several ambitious infrastructure projects — megacity construction, high-speed rail networks, not to mention the country's much-vaunted Belt and Road Initiative. But these won't reshape history like China's digital infrastructure, which could shift the balance of power between the individual and the state worldwide."

Amy asked, "But I thought most perceptions were that

America was ahead with its development of Artificial Intelligence?"

John continued, "You are right, Amy. Despite China's considerable strides, industry analysts expect America to keep its current AI lead for another decade at least. But this is cold comfort: China is already developing powerful new surveillance tools and exporting them to dozens of the world's actual and would-be autocracies. Over the next few years, those technologies will be refined and integrated into all-encompassing surveillance systems that dictators can plug and play."

Grace added, "The emergence of an AI-powered authoritarian bloc led by China could warp the geopolitics of this century. It could prevent billions of people, across large swaths of the globe, from ever securing any measure of political freedom. And whatever the pretensions of American policy makers, only China's citizens can stop it."

John flipped up a line drawing of an attractive Chinese female holding a single flower. He continued, "When Mao Zedong took over, he arranged cities into grids, making each square its own work unit, where local spies kept "sharp eyes" out for counterrevolutionary behaviour, no matter how trivial. The poster invites people with sharp eyes to keep an eye out for anything unusual. And during the initial coronavirus outbreak, Chinese social-media apps promoted hotlines where people could report those suspected of hiding symptoms."

John continued, "The Xue Liang rural surveillance

program, which translates as 'Sharp Eyes,' and the urban 'Skynet' and 'Safe Cities' networks are ostensibly being deployed to defend the civilian population from threats of crime, terror and natural disasters."

Grace added, "About 200 million video surveillance cameras monitor China's streets, buildings and public spaces compared with 60 million in the U.S."

John continued, "And with AI, Xi can build history's most oppressive authoritarian apparatus, without using the manpower Mao needed to keep information about dissent flowing to a single, centralised node.

John showed a chart of several suppliers, "In China's most prominent AI start-ups—SenseTime, CloudWalk, Megvii, Hikvision, iFlytek, Meiya Pico—Xi has found willing commercial partners."

John continued, "And in Xinjiang's Muslim minority, he has found his test population. As Bob Ranzino would coldly describe it, a population for experiments. Today, in China's single-party political system, religion is an alternative source of ultimate authority, which means it must be co-opted or destroyed."

Grace added, "Xi cracked down, directing Xinjiang's provincial government to destroy mosques and reduce Uighur neighbourhoods to rubble. More than 1 million Uighurs were disappeared into concentration camps. Many were tortured and made to perform slave labour."

Grace continued, "Uighurs who were spared the camps now make up the most intensely surveilled population on Earth. Not all the surveillance is digital. The Chinese government has creepily moved thousands of Han Chinese 'big brothers and sisters' into homes in Xinjiang's ancient Silk Road cities, to monitor Uighurs' forced assimilation to mainstream Chinese culture."

I thought back to what Pastor Oscar had described in my first day in Geneva. How the Reformation's religious war had created a series of bloody massacres through France and Switzerland in the 1600s. Some situations don't seem to change.

John continued, "Thank you, Grace. Meanwhile, AI-powered sensors lurk everywhere, including in Uighurs' purses and trouser pockets. According to the anthropologist Darren Byler, some Uighurs buried their mobile phones containing Islamic materials, or even froze their data cards into dumplings for safekeeping, when Xi's campaign of cultural erasure reached full tilt. But police have since forced them to install nanny apps on their new phones. The apps use algorithms to hunt for "ideological viruses" day and night. They can scan chat logs for Quran verses, and look for Arabic script in memes and other image files."

Bigsy asked, "But what about installing a Virtual Private Network as a workaround?"

John Answered, "Uighurs can't use the usual work-arounds. Installing a VPN would likely invite an investigation, so they can't download WhatsApp or any other prohibited encrypted-chat software. Purchasing prayer rugs online, storing digital copies of Muslim books, and downloading sermons from a favorite imam are all risky activities. If a Uighur were to use WeChat's payment system to donate to a mosque, authorities might take note."

John explained, "The nanny apps work in tandem with the police, who spot-check phones at checkpoints, scrolling through recent calls and texts. Even an innocent digital association—being in a group text with a recent mosque attendee, for instance—could result in detention. Staying off social media altogether is no solution because digital inactivity itself can raise suspicions."

"Damned if you do and damned if you don't?" said Jake.

John continued, "The police are required to note when Uighurs deviate from any of their normal behaviour patterns. Their database wants to know if Uighurs start leaving their home through the back door instead of the front. It wants to know if they spend less time talking to neighbours than they used to. Electricity use is monitored by an algorithm for unusual use, which could indicate an unregistered resident. It is intense automated scrutiny."

Christina added, "In Russia they would still do most of this old school. With people. There's a lot more cyber technology in use now, but for Russia it's pointing outwards. In China it seems to point inward."

John added, "Uighurs can travel only a few blocks before encountering a checkpoint outfitted with one of Xinjiang's hundreds of thousands of surveillance cameras. When Uighurs reach the edge of their neighbourhood, an automated system takes note. The same system tracks them as they move through smaller checkpoints, at banks, parks, and schools. When they pump gas, the system can determine whether they are the car's owner. At the city's perimeter, they're forced to exit their cars, so their face and ID card can be scanned again."

"That's some totalitarian state," observed Clare.

John continued, "Absolutely, Xi seems to have used Xinjiang as a laboratory to fine-tune the sensory and analytical powers of this new digital embrace before expanding its reach across the mainland. The state-owned company that built much of Xinjiang's surveillance system, now boasts of pilot projects in Zhejiang, Guangdong, and Shenzhen. These lay "a robust foundation for a nationwide rollout," according to the company, and they represent only one piece of China's coalescing mega-network of human-monitoring technology."

"So, this roll-out would be like implementing the Precognition envisaged in that Tom Cruise movie?"

Grace noted, "China is an ideal setting for an experiment in total surveillance. Its population is extremely online. The country is home to more than 1 billion mobile phones, all chock-full of sophisticated sensors. Each one logs search-engine queries, websites visited, and mobile payments, which are ubiquitous. All of these data points can be time-stamped and geo-tagged.

Kyle added, "And because a new regulation requires telecom firms to scan the face of anyone who signs up for cell phone services, a phone's data can now be attached to a specific person's face. SenseTime, which helped build Xinjiang's surveillance state, recently bragged that its software can even identify people wearing masks."

John continued, "Until recently, it was difficult to imagine how China could integrate all of these data into a single surveillance system, but no longer. A cybersecurity activist hacked into a facial-recognition system that appeared to be connected to the government and was synthesising a surprising combination of data streams. The system could detect Uighurs by their ethnic features, and it could tell whether people's eyes or mouth were open, whether they were smiling, whether they had a beard, and whether they were wearing sunglasses. It logged the date, time, and serial numbers—all traceable to individual users—of Wi-Fi-enabled phones that passed within its reach. It was hosted by Alibaba and referred to Urban Wisdom, an AI-powered software platform that China's government has tasked the company with building."

John continued, "Urban Wisdom is, as the name suggests, an automated nerve centre, capable of synthesizing data streams from a multitude of sensors distributed throughout an urban environment. Many of its proposed uses are benign technocratic functions. Its algorithms could, for instance, count people and cars, to help with red-light timing and subway-line planning. Data from sensor-laden trash cans could make waste pickup timelier and more efficient."

John continued, "But Urban Wisdom and its successor technologies will also enable new forms of integrated surveillance. Some of these will enjoy broad public support: Urban Wisdom could be trained to spot lost children, or luggage abandoned by tourists or terrorists. It could flag loiterers, or homeless people, or rioters. Anyone in any type of danger could summon help by waving a hand in a distinctive way that would be instantly recognised by ever-vigilant computer vision. Earpiece-wearing police officers could be directed to the scene by an AI voice assistant."

"But don't the citizens have data protection rights?" asked Clare.

Grace answered, "Not as you or I know it. We might think of the Data Protection Act and GDPR as burdensome, but they provide at least some basic protection from computer scanning of our personal backgrounds."

John continued, "Urban Wisdom originally pitched itself to be especially useful in a pandemic. One of Alibaba's sister companies created the app that color-coded citizens' disease risk, while silently sending their health and travel data to police.

John added, "As Beijing's outbreak spread, some malls and restaurants in the city began scanning potential customers' phones, pulling data from mobile carriers to see whether they'd recently travelled. Mobile carriers also sent municipal governments lists of people who had come to their city from Wuhan, where the coronavirus was first detected. And Chinese AI companies began making networked facial-recognition helmets for police, with built-in infrared fever detectors, capable of sending data to the government. Urban Wisdom could automate these processes or integrate its data streams.

Kyle said, "But even China's most complex AI systems are still brittle. Urban Wisdom hasn't yet fully integrated its range of surveillance capabilities, and its ancestor systems have suffered some embarrassing performance issues: For example, the government's AI-powered cameras mistook a face on the side of a city bus for a jaywalker. But the software is getting better, and there's no technical reason it can't be implemented on a mass scale.

Jake said, "Of course, the fact that the AI is seeking an individual for jaywalking is itself of concern."

Kyle added, "The data streams that could be fed into a Urban Wisdom–like systems are essentially unlimited. In addition to footage from the 1.9 million facial-recognition cameras that the Chinese telecom firm China Tower is installing in cooperation with SenseTime, Urban Wisdom could absorb feeds from cameras fastened to lampposts and hanging above street corners.

John added, "It could make use of the cameras that Chinese police hide in traffic cones, and those strapped to officers, both uniformed and plainclothes. The state could force retailers to provide data from in-store cameras, which can now detect the direction of your gaze across a shelf, and which could soon see around corners by reading shadows. Precious little public space would be unwatched."

Chuck commented, "America's police departments have

begun to avail themselves of footage from Amazon's home-security cameras. In their more innocent applications, these cameras adorn doorbells, but many are also aimed at neighbours' houses. China's government could harvest footage from equivalent Chinese products. They could tap into the cameras attached to ride-share cars, or the self-driving vehicles that may soon replace them: Automated vehicles will be covered in a whole host of sensors, including some that will take in information much richer than 2-D video."

Kyle added, "Data from a massive fleet of these vehicles could be stitched together, and supplemented by other Urban Wisdom streams, to produce a 3-D model of the city that's updated second by second. Each refresh could log every human's location within the model. Such a system would make unidentified faces a priority, perhaps by sending drone swarms to secure a positive ID."

"Some of this is still fanciful, right?" asked Clare.

John answered, "All of these time-synced feeds of on-the-ground data could be supplemented by footage from drones, whose gigapixel cameras can record whole cityscapes in the crystalline detail that allows for license-plate reading and gait recognition.

"'Spy bird' drones already swoop and circle above Chinese cities, disguised as doves. Urban Wisdom's feeds could be synthesised with data from systems in other urban areas, to form a multidimensional, real-time account of nearly all human activity within China. Tracking every moment of every Chinese person's life."

John paused, then bowed, "I've finished," he said, "And I can tell from your remarks that this was sinking in."

"Now imagine fusing the three sessions together - Bob Ranzino's, SinoTech and mine. You can start to see the potential interest by the Chinese in Brant's clever AI system."

"Although," he paused and scanned the faces, "My sources tell me that it doesn't really work? - That was Levi Spillmann's original design, wasn't it? I heard it ran too slow to be usable?"

No-one's face changed at this question.

"I knew it, " said John, "You have all got your poker faces on."

"But don't worry," he smiled, "Your secret is safe with me!"

"And you know something? You should really invite me along to your session with SinoTech. I'll be able to keep an eye on proceedings - I speak Mandarin. And I know their moves."

Christina looked over to Chuck. He nodded very slightly. Amy looked first at Juliette, then at Rolf and Schmiddi.

Rolf spoke, "I think it would be a very good idea if you were able to attend the session with SinoTech."

Street Life

I play the streetlife, because there's no place I can go
Streetlife, it's the only life I know
Streetlife, and there's a thousand parts to play
Streetlife, until you play your life away

You let the people see, just who you wanna be
And every night you shine, just like a super star
That's how the life is played - a ten cent masquerade

Will Jennings / Joe Sample {Randy Crawford}

Beach Life

After such a day of head-filling information, I'd almost forgotten that we were going to a Beach Bar in the evening.

Chuck had done brilliantly well. When we arrived at the bar, we were pleased to see it was literally on the beach. The tables were casual, and we were able to push a few together and to crowd around.

Grace, Chuck, Christina, Jake, Bigsy, Kyle, Tyler, Clare, and I were all present.

"So, where's the rest of the gang?" I asked.

"Oh, I arranged for this to be a meeting in two parts. The Brant gang arrive at 19:30," explained Chuck, "And Amanda has had to go back to London."

"It gives us a chance to talk through the intelligence from Grace beforehand."

Grace spoke next, she had told Amanda already and was

passing the news around the rest of us.

Grace started, "Amusingly, I found the link about Qiu Zhang on LinkMe - the professional recruitment and contacts platform. A Chinese recruiter had Qiu Zhang listed as a contact. The recruiter, according to LinkMe, worked for a think tank in China, where Qiu Zhang had been based for part of her career."

"Grace continued, "It was an obvious hook being dangled by the Chinese recruiter, to pick up western Mandarin speakers and to offer them riches in return for information. And it turns out that once we knew, we even tested the link."

Grace added, "GCHQ alerted MI5 and an MI5 agent was dispatched to meet the recruiter in Shanghai. The agent confirmed that the recruiter was not a think-tank representative, but a Chinese intelligence officer."

"The recruiter offer was $25,000 for handing over government secrets."

Chuck interrupted, "Whoa? That was a bit of a blunt instrument approach?"

Christina answered, "Not really - the Russians would do the same. Make it enough to make it tempting for the target. The amount can always be adjusted later and even be used to keep the target hooked."

Grace continued, "Well, the MI5 agent had been given a digital memory card containing eight fake secret and fake top-secret documents manufactured by GCHQ that had details of a made-up spying operation. Once they had proved that the recruiter was real, by the simple expedient of getting paid for the secrets, MI5 removed their agent but kept the recruiter running, as a way to identify who else was transacting deals."

Grace continued, "That's how they had found Qiu Zhang. She was a regular feature on the email to the recruiter and it seem highly likely that Qiu Zhang was involved up to her neck in whatever was happening. We think Qui Zhang was high in the MSS hierarchy. We estimate her to be a Case Officer in the Operations Department of the Foreign Nationals Division."

Chuck smiled, "So she would be responsible for acquiring technology not cleared for export? No wonder she is humming around Brant's AI Research. They say that China's most productive way of acquiring foreign technology is by sending scientists overseas on scholarly exchange programs. It is how the Chinese first acquired the neutron bomb; they acquired the technology from the Lawrence Livermore Labs in California."

Grace continued, "If that story of the dangling recruiter was unique, it would just be an example of a low-level threat. But in the past year, two real former U.S. intelligence officers pleaded guilty to espionage-related charges involving China. The same system of recruitment was being used. They became an alarming warning to the U.S. intelligence community, which sees China in the same tier as Russia as America's top espionage threat. And remember, many of these espionage cases don't go public."

Chuck agreed, "These cases provide a glimpse of the growing intelligence war that is playing out in the shadows of the U.S.-China struggle for global dominance, and of the aggressiveness and skilfulness with which China is waging it."

Grace continued, "As China advances economically and technologically, its spy services are keeping pace: Their intelligence officers are more sophisticated, the tools at their disposal are more powerful, and they are engaged in what appears to be an intensifying array of espionage operations that have their American counterparts on the defensive. And now, it would seem, Qiu Zhang had been targeted with getting some technology from Brant which could only assist their intelligence efforts."

Christina spoke, "I think I know something about this. Russia is usually more careful than China. China simply plays the numbers game."

She added, "While some of Russian intelligence efforts, such as election interference, are loud and aggressive and seemingly unconcerned with being discovered, Russians are careful and targeted when trying to turn a well-placed asset. Russia has veteran intelligence operatives make contact in person and proceed with care and patience.

"Their worst-case scenario is getting caught," Christina said, "Russia takes pride in their HUMINT operations. They're very targeted. They take extra time to increase the percentage of success. Whereas the Chinese don't care."

"What you have is an intelligence officer sitting in Beijing," Kyle said. "And he can send out 30,000 emails a day. And if he gets 300 replies, that's a high-yield, low-risk intelligence operation."

Christina added, "Often, Chinese spies don't even look too hard. Many of those who have left U.S. intelligence jobs reveal on their LinkMe profiles which agencies they worked for and the countries and topics on which they focused. If they still have a government clearance, they might advertise that too. As a classic target, find a 52-year-old. They have kids in college and need extra money."

Christina continued, "So what do the Chinese do? Chinese intelligence operatives pose online as Chinese professors, think-tank experts, or executives. They usually propose a trip to China as a business opportunity. Especially to those high-rankers who have retired from the CIA, DIA, and are now contractors—they must make the bucks," Christina said. "And a lot of times that's in China. And they get compromised."

Christina added, "It's quite like the Cold War. Once a target is in China, Chinese operatives might try to get the person to start passing over sensitive information in degrees. The first request could be for information that doesn't seem like a big deal. But by then the trap is set. "When they hand over that first envelope, it's being photographed. And then they can blackmail you. And then you're being sucked in," Christina said. "One document becomes 10 documents becomes 15 documents. And then you must rationalise that in your mind: I am not a spy, because they're forcing me to do this."

Grace continued, "Christina is right. Two decades ago, Chinese intelligence officers were largely seen as relatively amateurish, even sloppy. Usually, their English was poor. They were clumsy. They used predictable covers. Chinese military intelligence officers masquerading as civilians often failed to hide a military bearing and could come across as almost laughably uptight. Typically, their main targets tended to be of Chinese descent. In recent years, however, Chinese intelligence officers have become more sophisticated— they can come across as suave, personable, even genteel. Their manners can be fluid. Their English is usually good."

"So, is that how it seems with Qiu Zhang?" asked Jake, looking at me.

"Do you know something? I really don't know. I'm too far down the pecking order to have met the very senior Qiu Zhang," I said.

"Okay, we'll ask the others, when they arrive," said Jake.

Breakout

Wednesday, and it was almost time for the appointment with the people from SinoTech. We were using the breakout area instead of the Executive Briefing Suite because the breakout was where Rolf and Hermann had arranged the extensive and theatrically over-engineered demonstration rig.

All of us that worked with the equipment day-to-day knew it would perform too slowly and we wondered what reaction that would create among the Chinese delegation.

Rolf had set up a relay from the breakout area to the Briefing Suite, and Grace, Kyle, Tyler, Bigsy and Xuan Yahui watched from the comfort of the Hotel.

We'd had to decide who to place in the Demo area and had Amy, Juliette, Rolf, Hermann, and myself as the guinea pig - or should I say rat? Bob was present, as was Jasmine, and we assumed this was to add some gravitas.

We'd also brought along the backup in the form of

Christina and Chuck, both of whom knew how to manage unexpected developments.

Rolf had set up the Cyclone and added in the interface that Tyler had brought from Amsterdam. Juliette was acting as the animal wrangler and had the usual two rats, a white one for the start of the experiment and then a larger black one to be introduced part way through. There was also a large stash of chocolate buttons, which we knew made the rats go mad with interest.

Rolf handed over to Bob to introduce the session and I noticed that he flailed around without any form of auto prompt. I was silently quite pleased to see him look just a tiny bit foolish, although I was sure it would be bounced back onto one of us after the delegates and left.

Then it was Hermann's turn to describe the experiment, which he did in perfect English. I notice his English had suddenly become better than his usual pronunciations.

Rolf moved forward and Juliette placed the white rat carefully into the cage area. I felt, as usual, like I was stepping off a cliff and was soon inside the machine again. There seemed to be differences, and I wondered if they were caused by Rolf's latest adaptations.

Most noticeable was the lack of bright light flashes. I could also hear the Chinese delegates talking to one another. In fact, they were overriding the normal signals that I'd get from the rat. I tried to move the rat toward the chocolate buttons, but it was the usual laboured process.

"He's moving very slowly," said one of the Chinese to

another.

"Yes, at this rate, the rumours must be true. They have something, but it is too slow to be useful."

"Can it be speeded up?"

"I doubt it and look at the exertion, That man is sweating, just to move the rat around."

I suddenly realised. The Chinese were speaking Mandarin, but somehow, I could interpret what they were saying. I realised it must be the interface unit and its connection.

Rolf and Juliette were oblivious to this and were introducing the second black rat into the experiment area.

Then, a strange thing happened. I feel a resonance from the black rat. It was signalling to me! 'Don't fight about the chocolate. There's enough for us both, we can store it.'

It walked up to my face and looked at me with its eyes. I realised there was no flashing pattern around it.

"Why is it so slow," asked one of the delegates, in Mandarin, to the translator.

I almost answered directly, but Rolf answered, in English. "We have everything set to maximum."

"This isn't a good system, this is no diamond with a flaw

it is a flawless pebble," said one of the delegates,

"We can't use this," said another, "There's no point in inviting them to Beijing. Our own system runs better than this. The rumours were right. They have built a slow idiot of a system. A bird does not sing because he has the answer to something, he sings because he has a song."

I was very surprised that I was keeping up with their spoken language. I assumed it was something to do with Rolf's new interface and the adapter that Tyler had brought from Amsterdam.

Suddenly I felt myself tipping forward and the familiar Breakout Room returned to its normal view. I'd been disconnected by Rolf.

"Heart rate was running at around 200 bpm," he explained, "past the safety cut-out."

After such a day of head-filling information, I'd almost forgotten that we were going to a Beach Bar in the evening. I absent-mindedly put my hand on my shirt and could feel the dampness. I'd overheated again.

The Chinese were continuing to chatter, but now it was unintelligible to me, even although I was still wearing the Cyclone headgear.

"Thank you," said the Translator, "That was a very interesting demonstration and shows us just how far you have got with your trials. We from SinoTech wish you good fortune as you continue to develop your systems. We will be in touch to discuss the possibility of a visit to

Beijing over the next couple of weeks."

They were preparing to kick the Createl system into the long grass. Bob and the others would be most unhappy. At least Bob had been present for the demonstration, although I had a feeling that he, with his drive-by handshake, was preparing someone else to be the scapegoat. Project Management 101: Blame Avoidance and Credit Piracy. It seemed to sum up Bob's attitude, aided and abetted by the disarming Jasmine. They both had access to too much power.

The delegation was preparing to leave. Bob stood across the doorway, but the Chinese translator and a couple of the others from the delegation simply said, "Thank you and goodbye" in a disarmingly informal way. Everyone trooped out, leaving Bob to flail.

I could see Hermann and Rolf looking unperturbed by what had just happened. They both knew that this demonstration had probably gone further than any of the preceding ones. They had warned that the system simply wasn't ready.

Bob spoke, "Can we all - the Brant people - sit back down again. I have a few things to say."

Chuck and Christina kept walking toward the exit. I guess they had seen it all before.

Bob selects a scapegoat

Bob spoke, "I repeat - Can we all - the Brant people - sit back down again. I have a few things to say."

He seemed oblivious to the fact that two of the 'Brant' people had already left.

"You all told me that this system was ready for demonstration. That it would blow our clients' minds with just how good it was. I didn't see that. I didn't see any evidence of a superlative solution. I just saw a restless delegation, looking like they wanted to leave. It just wasn't good enough."

"No - I disagree," said Rolf. "It was the best and most forward-looking demonstration we have ever done. I'm amazed that you think it was not good."

"No - Rolf? Is that your name? I can't agree with you. I was not wrong. You must have lied to me about your progress. Hell, you are even lying now about how much progress we made. Can't you see it with your own eyes?"

"I promised those people from SinoTech that we had something great to show them. We didn't deliver. I expected they would see something that would make it a no-brainer for us to go to Beijing, present it to their bosses and create a huge sale for Brant."

"And you, Amy, are no better. I've already had to remove Kjeld for under-delivering, next it will be you. You should keep those people in the Lab under control."

Amy replied, "Let's face it, Bob, you took over the presentation at the last minute. You didn't tell anyone what you were doing, and you switched that session from one about our latest research work to one that self-aggrandised your own position at the expense of others."

"Amy, Out, Leave this meeting. Now,"

"You can't talk to me like that," said Amy.

"Oh, trust me, I can. Jasmine? Can you make some notes as I formally discipline Amy now, in front of her colleagues."

Jasmine stepped forward. She held up her phone, "It's okay Bob, I'm recording everything you say."

"I wish there were more like you, Jasmine," said Bob, "You are good at anticipating the next move, unlike Amy and her team,"

Bob continued, "I wanted to make this a great victory for me, so that we could show that I had created the exact system to propel Brant forward and make the company a lot of money. I first positioned it during the Company's Keynote. My speech received the best scores."

I was wondering how this had become 'all about Bob', like he was having a meltdown right in front of us.

"And as for you two," he glared toward Rolf and Hermann, "You can't just swan onto a plane and treat it as an extended drinking vacation. Jasmine tells me you've been out to Las Ramblas every night."

"Just a minute," said Hermann, "We invited Jasmine along with us one evening, because she said she didn't know anyone here."

Rolf glowered over towards Jasmine, who sat pretty but inert, very much like she had done that first time I met her.

I'd decided that Bob was having a tantrum now, and that nothing we could say would salvage the situation.

"I'm going now," I said, "I'm going to find a bar on Las Ramblas and I'll sit there with a jug of Sangria and some Tapas."

I stood and walked toward the exit from the Breakout Area. Outside, I sat for a few minutes until first Hermann and then Rolf and Amy together walked out.

Bucket of beers

"We should go back to the Exec Briefing room at the hotel and have a fuller discussion of this," said Amy.

We all nodded our heads in agreement. I could see that Amy was fairly seething from what she had just experienced from Bob.

We turned the corner from the exit from the Exposition Hall and walked back along the short looping drive and into the hotel.

The first thing we noticed, once again, were the black outfits of a couple of Chinese Policemen standing just inside the revolving doors. They both wore dark glasses and the outsized hats.

"Nothing like being obvious!" said Hermann as we looked toward them.

We made our way to the glass lifts and up to a high floor. There we sought our Briefing Suite and entered it. There was an outer annex with refreshments and a waiting area, and then another room with a large oval table. The windows looked out onto the fields behind the hotel and in a distant one I could see a couple of horses running back and forth in the sunlight.

"Bob seemed to lose it at the end!" said Kyle, addressing us all as we entered.

"We've just ordered some room service to be sent up. Coffee and water, would any of you like anything else?"

"Maybe some sangria?" said Hermann. I was impressed at his robust ability to keep on an even keel through all the nonsense.

Amy nodded, "Yes, and a gin and tonic. Maybe a double."

"And a bucket of beers," answered Rolf.

"Now we're talking, " said Bigsy.

It was John who described what the Chinese had been saying to one another, " They said the demo sucked. That it ran too slowly. They were unimpressed and didn't think it would be worth taking the test rig to Beijing."

I had to decide how much to say about what I'd experienced in the test rig. We were a large group now, and there was a risk that the word would get out. After all, the Chinese had already picked up that they thought our test rig was faulty. And specifically, that they thought it was too slow.

I weighed up my options. Kyle and Tyler were leaving the next day, and of course I knew them both across many years. Now, in Geneva, I worked closely with Rolf and Hermann, Amy, and Juliette.

Christina and Chuck seemed to have my interests at heart. They would also vouch for Jake, Clare and Bigsy.

Grace and Amanda both had previous good credentials, and I'd continue to trust them both. I seem to have everyone covered, except possibly John.

"John, I'm so glad you pre-briefed us about the Chinese situation and SinoTech. How much longer will you be here in Barcelona?" I asked.

"I'm leaving tonight, as a matter of fact. I could only get a short time at the Symposium, but I think I've been to all the best bits!" John answered, "I wouldn't have missed this for the world."

I decided that my news from the experiment could wait until we'd slimmed the numbers by three people.

Instead, we talked about the news that we'd received. "So, Amy, did you know that Kjeld had gone?" asked Hermann.

"I had no idea, until Bob said it," said Amy, "Let alone why he has gone, or where for that matter."

"Could it be the terrible speech he'd written?" asked Grace, "You remember, the one I got a part of, in advance?"

"Maybe, but I suspect there's the pretty hand of Jasmine in this somewhere," answered Amy, "She may act a bit ditzy, but she seems to carry a big stick."

"Yes," said Christina, "My thoughts as well, she acts like a missioned agent; someone who has to clear a path for others. I'd scream though if I had to do what I know she is doing!"

I noticed Bigsy fiddling around with his bag of technology. He produced the scanner and signalled to everyone in the room to be quiet. He flipped it on and walked around silently checking for bugs.

"Sorry everyone, I suddenly realised that this event is so well equipped to bug us, that we'd better take precautions. We're in the clear, though. I've only found our own cabling and technology."

I wondered if we'd all been subjected to too many traumatic situations over the last few days and it was making us all jumpy.

"W", said Chuck, "Let's go to the W. It's on the beach, at the far end of the town. 20 minutes by taxi and we'll be in a luxury beach resort. Bring swimwear."

"Now you're talking," said Amy.

"I'm gonna bail," said Kyle, "I'd love to, but I've got to leave early tomorrow morning."

"Me too," said Tyler, "I guess I'm on the same flight."

"And me, " said John, "I'm leaving tonight. Already packed and checked out, actually."

I suddenly realised what Chuck had achieved. He'd pulled together the core set of people for the ongoing discussion.

W

We bundled into the taxis to the W.

Chuck got it right again. We were soon in a sun-soaked resort hotel, with pools and spas in all directions. Everyone went swimming and then, once recalibrated in the evening sunshine, we gathered around a fire pit and I told them about my Cyclone experiences.

"Rolf's new gadget worked," I said, "I could understand the Chinese guys speaking. There was lag, but it was totally obvious what they were saying. And the same with the white rat. Not speech, but I could communicate with the other black rat on a perfunctory level. We agreed not to fight and to share the chocolate buttons. I just knew that was what the black rat wanted. There was also much less of the flashing lights interference. I think I'm getting acclimated to the Cyclone."

"So you could tell what the Chinese were saying?" asked Amy, "That's a major breakthrough!"

"And the delays that appear across the whole system

don't seem to affect the plug-in modules?" asked Rolf.

"There was still some delay, but nowhere near as much as when I tried to control the rat, " I said.

"Remember, we don't have the full system here," said Hermann.

"What do you mean?" asked Amy.

"Oh, we had to leave some components back in the lab. Otherwise, it would have been too complex to bring everything here," lied Hermann, realising he'd said slightly too much in front of Amy.

"No matter," said Amy, "But I'm glad that it turned out without us having something to donate to the Chinese."

"Donate?" I asked.

"Well sell, then," answered Amy, "Although it's a stock trade that mainly affects those that already have wealth. For the rest of us it is no different except that some of our bosses will change."

"But it would change the very balance of power on the planet if Brant sold this technology to the Chinese!" said Chuck, "You can see what they are already doing. Imagine surveillance enabled systems leaching out of China into global households. Imagine AI-equipped policing systems in the wrong hands. It's a truly dystopian outlook."

Rolf nodded, "Yes, it is better that we never find the

streamlined version of the Createl coding. Unless it was an emergency."

"I'm guessing that is what Levi planned," said Juliette, "To stop the system from ever being commercialised."

Barbecued food arrived.

"This is heavenly," said Amy, "Good call, Chuck."

We all stared out into the Mediterranean as the sun began to set.

"What's the next part of the plan?" asked Jake.

Clare pulled out some paper and some marker pens.

Thursday Split

Thursday and it was time for us to split up. The organisers ran the conference through to Thursday, but it was partly so that could host a big flamenco party event on Wednesday evening. The event seemed more subdued on Thursday morning and I realised we had probably made a good call on Wednesday to go to the W to relax poolside.

Rolf and Hermann were staying on for an extra day to disassemble the hardware, whilst Juliette and Amy and I were all heading back to Geneva. I had a sneaking suspicion that Rolf and Hermann could uninstall the kit in about half an hour and would then have the rest of the day and evening to hit the bars of Barcelona.

Chuck, Christina, Jake and Clare were all heading back to London, but Bigsy had said he wanted to come back to Geneva with Juliette, Amy and me. He said he had some further tests to run and a working theory about how Levi's system worked.

Grace had left early in the morning and was on her way

back to Bristol and then to Cheltenham.

We'd all received a message from Amanda to connect into a call Thursday evening at 19:00 UK time, to cross check our findings. Amanda said she was on to something new.

Tektorize

Later, I'd completed the sweep, unpacked, and was just filing things into those 'return from a trip' heaps. Clothes (clean), clothes (used), toiletries for bathroom, gadgets, chargers and cables, papers - when my phone rang. It was Juliette.

"Oh Hi! I wasn't expecting to hear from you so quickly."

"We've been robbed. Well - the Lab has been robbed. They've taken the Number 2 system - Amy called to tell me. She said that two French trucks turned up and several men speaking French had asked to remove the system, and that we needed it in Barcelona."

"They had documentation and shipping instructions, so the Lab simply gave them everything. They had forged Amy's credentials onto the paperwork, but it was good enough to fool the security people."

"Look, I'm sending you across the paperwork now. Security said the guys all looked and sounded French. They seemed to be hired from Prévessin-Moëns, France.

There was just one unusual thing.

"On the back of one of the dockets someone had written something in Chinese Hanzi ideograms. It was a take-away order from Hungky. That's just about the best Chinese restaurant in Geneva. I've sent you the photo of the document:

饥饿：扬州炒饭，香酥鸭，盐和胡椒茄子，上海小笼包。

It read: Yangzhou fried rice, crispy duck, salt and pepper aubergine, Shanghai Xiaolongbao - it's an order from someone who knows their Chinese food, apparently!"

"Okay, so they have got the second system, but what about the one we hid? In the cupboard?"

"I didn't ask Amy because I didn't want to alert her to our deceptions. But only Herman, Rolf, you and I know about that system and that it contains Levi's key."

"And Bigsy," I added, "Bigsy knows as well."

I decide to call Bigsy and tell him the news.

Bigsy was surprised but then said, "I've got some for you as well. I received an email from your friend Kyle - I think you've got it too; he sent it to most of us.

Bigsy continued, "Remember Kyle sent a monitoring probe into the Chinese policemen's camera surveillance system? It turned out to be very interesting. The big units like the 'parking meter probes' were filming everyone, but the eGlasses that the police were wearing were more specific. I guess it was because of their capacity, but they were only looking for certain categories of person. They were looking for Brant personnel -that was anyone with Brant Exposition Badge. SinoTech - people from the SinoTech Corporation, all the way from workers to executives and - surprisingly people from Tektor Issledovaniye - That's Tektorize, by the way."

"Tektorise?" I queried.

"Yes, Tektorize with a zed. A Russian outfit, based in Moscow and researching - guess what - Artificial Intelligence!"

"Okay Bigsy, had you heard of them before today?"

Bigsy answered, "Nope, I had to look them up, but they seem to be big in the Kremlin ever since Putin asked for a national AI strategy. Look, I've already passed this info on to Amanda, and she was going to alert Grace as well. I've received some other papers which I'll send on to you and I'll be around in a few minutes, see you!"

While I waited for Bigsy and the return of Juliette, I looked up Russian AI firms with Google and was genuinely surprised. There were dozens of them, but they all seemed to be quite small and their descriptions were quite bland and scattergun. An example: "Intelligent Buttons LLC - Digital boost for your business to succeed! An IT development company accelerating your digital transformation. We build cutting edge IT solutions integrating into your business strategy to help your company grow. A remote dedicated IT department perfectly matched for your requirements can successfully develop end-to-end IT products or add value to existing ones. Based on our 10+ years of expertise, we develop industry-specific solutions for Travel, Fintech, E-commerce, Medtech, Real Estate, and Networking."

I looked at the next few entries: Clever Machine Learning Corporation. Computer vision, Internet of Things. Your challenge is our inspiration — we build customer software solutions for the most challenging of projects. CML's custom software developers have been helping the world's top companies to turn their dreams into reality. Our focus: Machine Learning - recommendations, prediction, forecasting, classification - Computer Vision solutions - OCR, pattern recognition etc. Image Processing - Natural Language Processing & Text Analysis - syntax, semantics, sentiment analysis, computational linguistic.

And so it went on for another two dozen smaller companies. No mention of Tektorize. They mostly seemed like the type of companies who could join systems together – Systems Integrators - but not the kind that would be inventing new cutting-edge technology. They were all fishing for slightly dim clients. The sum of all dims.

I remembered when Kyle and I had made our illegal casino scanners. We were looking around at ideas. For amusement, we'd frivolously bought a Russian clock allegedly from the Soyuz program from an eBay auction and dismantled it. We were shocked that it contained about half a dozen printed circuit boards and many surface mounted Transistor-Transistor-Logic chips. The thirty-year-old gadget was supposed to be from the 80s but looked more like DIY project from about 1960, even if there was a 21st Century Haynes manual for Soyuz. The gap of twenty or thirty years reminded me of the difference in sophistication when one looked behind the scenes at some of the space technology. It appeared that the Russian AI was similarly behind the curve.

By now it was time for the call with Amanda. I was sure that she would have been briefed by now by Grace or someone from London. By this time, Bigsy was round at my Apartment, as was Juliette.

We all joined the call and Amanda jumped straight in.

"Hello everyone. We can make some sense of this now. I have checked with our agent in Peking University. We can all play the embedded agents game! Peking University is where Mary Ranzino worked when she was seconded from Harvard to work on the Chinese Thousand Talents Plan. It is also where we discovered that Qiu Zhang worked and that is how Qiu Zhang came to be working for Brant, via her connection to Mary."

"Our agent is a leading Chinese artificial intelligence researcher in the US who returned to China as Beijing strives to become a global leader in the field. Peking University said it was working with Beijing city and the central government to set up a new and separate AI research institute in collaboration with other leading Chinese universities. Grace, can you tell us more?"

Grace continued, "The University has been working on

the human brain's cerebral cortexes and their use for processing vision. And the University states that 'research into computer vision opens the gateway to artificial intelligence.'"

Amy interrupted, " You can see why the work at Brant would be of such interest, and of the reason that they would like to gain access to Levi's work."

Grace added, "Call this Item 1: The University published a research paper establishing the potential for robots powered by artificial intelligence to earn the trust of humans in a project funded by the US Defence Advanced Research Projects Agency (i.e. DARPA). It talked about 'explainable actions' - the robots could explain their reasons for doing things.

Chuck commented, "But wait. Here we see the Chinese trying to promote its findings to the American Department of Defence."

Grace continued again, "The same University has also employed an ex-head of artificial intelligence from Microsoft, so they are making some big statements about their direction."

"Well that's not all that I got from our agent," said Amanda, "He described their Qiànrù sh" zhōngguó - Embedded China Programme."

Grace continued, "We can call this Item 2: It was a tough one to find any more detail about. I finally found it described as 不稳定系统 - Bù wěnd"ng x"tǒng - Systems of Instability. A kind of lo-tech hustle."

She continued, "We've all heard about America attempting to stop the march of Chinese 5G systems into their critical national infrastructure? It turns out that there's an almost unstoppable force. Consumer electronics and the Internet of Things. Everyone wants everything to be controlled from their phones. Vacuum cleaners, washing machines, lights, heating, doorbells."

She paused and then added, "Well, it doesn't matter how American the name sounds, most of that equipment is built in China. Add a few cheap components - microphones, cameras - and these devices will positively sing!"

Grace added, "This is Item 3: Mapping of the areas scanned. A way to mark the exact coordinates of the data gathered. It's a combination of WIFI tracing , camera scanning and drone mapping. It provides a comprehensive inventory of an area."

Grace added, "And Item 4: Amanda mentioned that Bob Ranzino has gone. Well, he's only taken minutes to pop up elsewhere. He has turned up in Washington, D.C. at the building of Tektorize, on Pennsylvania Avenue NW, which is halfway between the White House and the FBI building and more-or-less opposite the Trump hotel. And he's staying in Trump's hotel with Jasmine Summers."

"I knew it," said Christina, "Jasmine Summers - an unusual name if I may say so - is being run by someone and acting as a fragrant path clearer for Bob Ranzino."

"Any more?" asked Clare, "Only this is getting complex. Here. I've summarised it."

Jake looked relieved as Clare screen shared her list.

1. Peking University as a hub trying to develop AI for Military purposes.
2. Chinese deployment of sensors into every consumer product.
3. Comprehensive mapping of citizens and their activities.
4. Bob Ranzino to Russian-run Tektorize, with support from Jasmine Summers.
5. Chinese State interest in Brant and SinoTech.
6. Theft of experimental Selexor and Createl system from Geneva - by Chinese?
7. Mary Ranzino still at Brant, run by Chinese including Qiu Zhang.
8. Murder of Levi Spillmann by Chinese agents (Bérénice Charbonnier, Niklaus Zeiler, A.N.Other?) under command of Qui Zhang.
9. Brant trying to hype a sale of the AI system, to raise their share price.

Clare looked triumphant as she flipped back from screen share.

"I'll email it to everyone."

The Genevan.

We all said our goodbyes in Barcelona Airport, and I flew back to Geneva. It seemed as if I'd been away for much longer than a week. Juliette gave Bigsy and me a lift back to his hotel and then onward to my apartment, but then went straight back to her place. Her car with Bigsy in the back seats, had been jam-packed with all our luggage.

I was picking up my mail from the letterbox at the entrance to the apartments, when I saw Bérénice, who waved. I made a mental note to re-sweep my apartment for listening devices when I got inside.

Among the mail was a newspaper - Le genevois - The Genevan. That was the paper that Bérénice worked for. It wasn't the full newspaper, just a subset that they published in English because of all the ex-pats living in Geneva. I noticed one of the headlines.

Not just CERN, we have other scientists too!
By our own correspondent

It looks as if there is going to be a scramble for the rights to Genevan product RightBrain, out of the Laboratories of Brant Industries.

Brant is offering the rights to RightBrain to the highest bidder, in return for an equity share swap. So far, the Chinese company SinoTech and a Russian corporation Tektorize have shown an interest, but the bidding war hasn't started.

Brant Industries is, of course, American, although the huge unit based in Switzerland is truly Swiss, with an international team of scientists. Expect to see plenty more scientists in Geneva as the going gets tough! And if you don't know what the mysterious Brant Industries do here in Geneva, here's a description.

About Brant
By our business correspondent

Brant Industries RightMind is an Artificial Intelligence company developing wearable brain–machine interfaces (BMIs). The company's headquarters is in Washington, D.C.

Since its founding, the company has hired several high-profile neuroscientists from various universities. It has received $200 million in funding from Venture Capital Funds and a further $100 million from Brant Industries. At that time, RightMind announced that it was working on a 'Cycle helmet like' device - The Cyclone - capable of working with the Createl System and linked to Selexor to provide Artificial Intelligence to augment human thought processes.

This has been demonstrated under lab conditions with a lab rat configured with more than 200 electrodes. RightMind anticipate starting experiments with humans in the next twelve-to-eighteen-month timeframe and predict to have fully functional augmented AI capabilities ready for human use within the next four years.

Brant's Technology

Our science correspondent writes:

The company has "remained highly secretive about its work since its launch", although public records showed that it had sought to open an animal testing facility in Washington, D.C.

It subsequently started to carry out research at the Brant Campus in Geneva, Switzerland.

Recently during a live presentation at the Barcelona Future Intelligence Expo, the RightMind team revealed to the public the technology of the first prototype they had been working on.

It is a system that uses the Cyclone headgear, connected to Brant Industries Createl AI system, supported by Raven Industries Selexor configurable AI integration platform.

Cyclone Probes

The Cyclone probes are composed mostly of light and magnetic sensors that locate electrical signals in the brain, and then provides a sensory area where the probe interacts with an electronic system that allows amplification and acquisition of the brain signal. Each of the probes contains 32 or 64 wires, each of which contains 32 independent electrodes, wired into the probe receptors in the Cyclone helmet.

Think of the outputs as so many signals lighting individual pixels on a camera screen. All the different colours can be displayed too, where all of the signals represent either human autonomous process like breathing, to human personality processes like anger and love. Literally shining lights representing brain activity.

Inductance

Studies involving the use of inductance probes with the brain have shown that, due to the brain's plasticity, it eventually acclimatises itself to the sensors such that can adapt its outputs to assist the sensors to function and complete their signalling.

Electronics

RightBrain has developed an Application-Specific Integrated Circuit (ASIC) to create a multi-channel recording and replay system. The system consists of large numbers of neural amplifiers capable of being individually programmed with analog-to-digital chip converters within the chip (ADACs) and a peripheral circuit controller (PCC) to serialize the digitised information obtained.

It aims to convert information obtained from neurons into an understandable binary code in order to achieve greater understanding of brain function and the ability to stimulate these neurons back.

Algorithmic 3D Neuron detection

Currently, the probes are still too big to record the firing of individual neurons, so they can record only the firing of a 3D group of neurons; RightBrain representatives believe this issue might get mitigated algorithmically, but the massive computational expense does not produce exact results. Instead, further iterations of research will look to speed up the computation through the refinement of better algorithms.

Criticism

At a live in-house demonstration one of their early devices was described as a healthcare breakthrough which could cure paralysis, deafness, blindness, and other disabilities.

Many neuroscientists and publications criticised these claims, describing them as described them as "highly speculative" and "neuroscience theatre".

Further critics suggest that the research path illustrated by RightMind was one which ultimately projected militaristic uses for the Cyclone and for RightMind.

Shining Light

Roman candles that burn in the night
Yeah you are a shining light
You lit a torch in the empty night
Yeah you are a shining light
Yeah you light up my life

Beneath a canopy of stars
I'd shed blood for you
The north star in the firmament
You shine the most bright
I've seen you draped in an electric veil
Shrouded in celestial light

We made a connection
A full-on chemical reaction
But by dark divine intervention
Yeah, you are a shining light

Timothy James Arthur Wheeler - Ash

Gather thoughts

Amanda spoke, "We can stop this, stop Brant from selling a weaponised AI to the highest bidder."

"How so?" asked Chuck.

"The Chinese already believe the Brant system comprising Createl and Cyclone doesn't work, from that session in Barcelona. We'll need to similarly persuade the Russians."

"What about the 'loose players'?" asked Chuck, "I mean - there's a whole group of them: For the Chinese there's Bérénice Charbonnier, Niklaus Zeiler, Qiu Zhang, Mary Ranzino and SinoTech and then for the Russians there's Bob Ranzino, Jasmine Summers plus Brant and Tektorize."

I could see that Christina was thinking.

"The Chinese hold all of the cards now. They know about the unreliable nature of the system. They have stolen one as well and will soon see its limitations. The question is, where will they have moved the stolen system?

"That's a good question!" said Amy, "After all, it is a lot of sensitive equipment. They could move it somewhere nearby, but I doubt whether they would take it out of the area, for fear of wrecking it in the process."

Clare asked, "Are there any other nearby Lab facilities where they could rebuild it?"

Grace cut back in, "Well, Geneva. Think about it, CERN. Conseil Européen pour la Recherche Nucléaire - The European Council for Nuclear Research. You know, the people with the huge particle accelerators."

"You make a good point," said Juliette, "And most people think of CERN as in Switzerland, but there's a large amount across the border in France. The name of the place? Prévessin-Moëns, France. That's the same as the place that those French vans came from.

"In Geneva at CERN there's all the prestigious buildings, but where CERN test big magnets and supercool liquids it is less glamorous. Most of the buildings look like elongated sheds. Maybe silver and blue army barracks."

"Well that is good to know, " said Bigsy, "Kyle, the security expert, showed me how to install a little extra code into the units - I'd almost forgotten! They send a homing signal across the internet. We should be able to track the system down with that, especially if we know the area we are looking for."

"Okay," said Christina, "So we can find the stolen system. We still need to use it for a demonstration to the Russians. Just like the demo in Barcelona, it shouldn't fill the Russians with joy."

Bigsy interrupted, "Hey - here we are! I've found the system. The finder is a bit like 'find my phone' and gives me a map showing the whereabouts of the system. Juliette, did you say that would be at Route A. Einstein? In Meyrin, Switzerland?"

"No, I think that road goes right across the border, you need to look further west and you'll be in France."

"Ah yes, I can see a series of concrete and corrugated buildings. It doesn't give me the impression of CERN though, more like an out-of-town factory estate."

"No, that will be CERN," answered Juliette, "The big facilities are all in Switzerland, but there's a lot more in France. Can you locate the system any more precisely?"

"I can see some other systems on the same network. They have names like AMS POCC and AMS SOC? The Wi-Fi Nodes are all for CERN_189."

Juliette answered, "You are near the centre of the Big Magnets section. It's the Alpha Magnetic Spectrometer - Payload Operations and Control Center and the Science Operations Center. That section of the site is all paid for by NASA and is to do with Space operations - Cosmic rays, I think. CERN_189 is the building number - but that's a pretty random designation."

"How do you know so much about all of this?" asked Clare.

"Before Levi, I dated a French boy Christian Côté - I must have a thing for scientists, I guess. He worked at the CERN facility."

"How long did you date?" asked Clare.

"Oh, not long, he was obsessed with cosmic rays. Even his favourite song was Shining Light by Ash - see the connection - and he said I reminded him of Charlotte Hatherley."

"Nooo. He didn't!" said Clare wincing.

"Who is Charlotte Hatherley?" asked Bigsy.

"It's not the point," said Clare, "It's like being told that you remind your boyfriend of his last girlfriend or his favourite crush."

"Oh, I see what you mean," said Bigsy.

"Charlotte Hatherley was the underrated lead guitarist for Ash," said Christina, "I met her once an age ago when I was in a studio. She was a hot kick-ass guitarist but sick of being asked by the press what it was like being on the road in a bus with 10 blokes."

Chuck spoke, "I think we have a way into CERN, beyond Juliette's acquaintance. When I was doing the missile testing, we used to need to know what would happen to guidance systems when they went into the exosphere. We used CERN to tell us the answers. I reckon I still know some of the people there. Does cern.ch still work?"

Bigsy laughed, "Yes, it does! It even gives the main switchboard and a bunch of numbers!"

"I expect MapCERN also works, then. You can probably

dial-up Building 189 on the map!" said Chuck.

A moment and Bigsy screen shared his computer.

"Here we are. A big tin shed on the outskirts of the facility. Just right for unloading the stolen equipment into a well-provisioned building that won't draw any attention!"

Grace said, "Right, so if the equipment is there, we really need to demonstrate it to the Russians! So that they see for themselves that it doesn't work."

Duncan Melship

The session continued, and I stepped away to make some coffee for the three of us.

I'd worked out that the still mysterious Christina must have a hot line to Russia, because she said she'd been in contact with someone in Moscow. She had spoken to a Russian diplomat Fyodor Kuznetsov to see if the Russians knew anything about RightMind.

What she got back startled me, although Amanda and Grace didn't seem in the least bit surprised. Apparently, Putin had demanded that The Ministry of Defense of the Russian Federation, together with the Ministry of Education and Science of the Russian Federation and the Russian Academy of Sciences, together tackle Artificial Intelligence: Problems and ways to solve them.

Christina had sent the papers from the original meeting to Amanda, Grace and Bigsy, and they were in the stuff that Bigsy forwarded to me, which I'd looked at while I waited for Bigsy and Juliette to arrive at the apartment,

To be honest, it all looked about as primitive as that Soyuz clock. The Russians seemed to be heading in the wrong direction with their complicated research although their own Conference in Lomonosov Moscow State University Biokyberkenetika had pictures of a wide range of scientists worthy of a Hollywood blockbuster central casting call. I flipped through the sessions and could see that they were still pursuing biomechanics as a basis for their ideas.

Some of it was what I call 'hole digging' - worthy but vertically structured into infinitesimally smaller detail about the Wrong Things. Someone needed to buy them a buzzer. Vroop. Wrong Question, keep moving, nothing to see here.

Christina spoke, "It turns out that the Russians are struggling to keep up with The Kremlin's demands. They are conducting a lot of research, but it isn't going anywhere useful. They are becoming increasingly desperate to gain useful data from the Americans and the Chinese. Oh, and I include Brant with the Americans."

Juliette spoke, "It reminds me of that session we arranged at the Lab. We ran a briefing a couple of weeks ago, with Professor Doctor Andreas Türkirchen, from the University of Zurich, who works in their Brain Research Institute. It was on message but gave us some room for thought of our own."

"That was my very first day in the Lab!" I said, "it seems an age away!"

Juliette continued, "The thing was, we were asked to put

on another presentation with it. It was by an oily British MP and was vacuous. The presenter was flanked a couple of metres either side with protection people, like he was a big shot, but his actual content was crushingly boring."

"кронштейн-защита - kronshteyn-zashchita - Bracket Protection," said Christina, "Think of two brackets enclosing an exclamation mark (!) It's a classic protection technique used by the GRU, when they are close marking someone. Additionally, a technique sometimes used is to substitute one of the close protection officers for someone else - such as a scientist wanting to gather information. Highly likely at that conference."

I spoke, "I missed the session, but the British MP wasn't even one that I've heard of: Duncan Melship, I think was his name."

"Yes, I had to speak to him because I was an organiser, but I found he is manner slightly creepy," said Juliette.

Clare chipped in, "Melship? I know his assistant, Lottie! I worked in Parliament for a while, assisting Amanda actually, and during that time I shared a flat with Lottie and her friend Tessa. Wild times. They had a nick-name for Melship...I can't remember it...something to do with his wandering hands."

"I can see the dots starting to form," said Jake, "If we assume that Duncan Melship is somehow tied up in all of this, then we could get to him via Lottie? Maybe even get a message back to the Russians?"

"Can we look up Melship, then?" asked Amanda.

Grace replied, "I was just doing so. The most striking thing is that Melship is on the Board of ISMC - That's International Strategic Management Consultants- They have previous form with GCHQ and SI6. Some of you will remember Gerhardt Schmidt and Sir Charles Frobisher, both from ISMC and it turned out that ISMC was being run by the Russian Mafia, but with a very respectable front to it."

Chuck answered, "That shake-down in London, at that Club - something to do with Raven Corp - that was linked to the Russians, I remember, although I don't remember anything about Melship?"

Amanda agreed, "That's right, we had both Gerhardt Schmidt and Sir Charles Frobisher implicated, but if you remember later, they both disappeared. Schmidt when his yacht was blown up, and Frobisher when he took a helicopter ride off the coast near Nice."

Christina said, "That's right, the assumed murder of Frobisher pointed to Vassily Turgenev, but Turgenev was later taken down by Anatoly Yaroslav working for Kasharin Timur Maximovich. That's Tima Maximovich, Head of Russian Infrastructure. And remember Turgenev, the enforcer, was a direct friend of Putin."

My own senses were reeling a little at all these sudden disclosures. It was as if the inconsequential-sounding Duncan Melship had been the key to unlock a giant portal. Several of the people on the call seemed to know a huge amount about Raven Corps, ISMC and the

Russian Mafia links. And of course, Brant Industries had taken over Createl but was a wholly owned subsidiary of Raven Corp.

"The obvious point is that this spells danger," said Chuck, "We've seen Maximovich in action before and he uses strong arm tactics to get what he wants. Yaroslav used a Surface to Air Missile against Turgenev and his gangster travelling companion Gavy Yegorin."

"It also points out what I was talking about the other day," said Christina, "When the old-guard Russian oligarchs get toppled by the young guns, with the Kremlin's quiet sponsorship. Instead of throwing them off balconies, nowadays it's out of helicopters."

Clare asked, "So what can we do? Get a message to Melship about the Chinese having the RightMind system? That they are storing it inside CERN premises!"

Chuck said, "It needs to be stronger than that. Maybe that the Chinese are selling the system to the highest bidder and that the American DoD is interested? That the only working version of the system is the one inside CERN and that the one used at the Exposition was a dummy rig."

Christina spoke, "That could work, but it needs to come out in a round-about way, not like we've just sent them a message in capital letters. They won't believe it unless they think they have got it clandestinely."

"Okay, I think I have a suggestion," said Clare, "Fortunately I now know how things work around

Parliament, Here's the plan..."

"...I'll ask Andrew Brading, the MP who I supported when I was a Research Assistant, to invite Duncan Melship for lunch. It can be in the Stranger's Dining Room because I'll say we want to bring a couple of extra people along. I can say it is a clumsy attempt at lobbying, and they want to get Melship to present at another AI event. Then we can take, say, Matt and Juliette, who know about the RightMind experiments and me plus Lottie as two surrogate assistants. I'd also suggest Christina as a form of protection officer for us all."

"But will Brading do this for you? I asked, "He's a pretty high-profile MP these days!"

"Trust me," said Clare, "I know he'll do it and I expect Melship will even feel a little flattered with the attention!"

"Can Lottie be trusted?" I asked,
    "Oh, totes!" came Clare's smiling reply.

Strangers' Dining Room

A couple of days later and Clare had visited Lottie. I was impressed that she'd punched a hole into Melship's calendar too and that the meeting in London was set for the following Monday.

Clare said that Lottie and Tessa were mad keen to see her and that they'd promised to keep things secret. I was now on a BA flight from Geneva to London, sitting next to Juliette and with Rolf in the row behind. Amy said it was better to show up with people who knew the detail of the system, for what we were about to attempt.

We'd arranged to stay over in the Park Plaza, which was literally just across Westminster Bridge from Parliament. Christina and Clare both knew it and would meet us there.

It was Monday, before the meeting and we were drinking coffee admiring the view towards Parliament when they arrived.

"Hey, Matt, Rolf, Juliette!" said Clare.

We all hugged one another and sat in a circle. Rolf was the most excited to be visiting inside Parliament and was hoping to get some selfies.

"We'll be meeting Melship in the Strangers' Dining Room, explained Clare. It's where MPs can invite members of the public to dine. It's quite nice, if a little bit old fashioned. I should say 'Traditional,' Rolf."

Clare told the story of her meeting with Lottie, "Lottie went berserk when I dropped around. She and Tess have been having a blast and they have a new girl in my room now. Her name is Hannah, and she seems to have settled in well - she is another assistant to a Conservative MP. Lottie and Tessa said they have had a quiet time compared with when I was there, and something seemed to happen every day. They still talk about 'The Pizza' that was delivered with a message - although they've changed it so that the 'secret' message was baked into the pizza instead of just a piece of paper stuck to the box."

Clare explained how she had got Andrew Brading to come at such short notice. She contacted Serena in Andrew's office and was overwhelmed with the praise and interest from her and from research assistant Maggie. They were both intrigued with her latest developments and said they were sure that Andrew would be delighted to host the lunch.

Clare described it, "They looked at his calendar and cleared everything away!"

Clare continued, "They even invited me to drop by to

Andrew's office an hour before the meeting so that everyone could catch up! I had to remind Serena that I no longer have a pass for Parliament! She said she would get me a Day Pass, but I would have to take the long way in."

Rolf asked, "So how do you know these people, Clare?"

"I worked inside Parliament for a while. I helped MP Andrew Brading when he was being attacked by the worst kind of lobbyists. Suffice to say I've kept good links with some of the people."

"Fantastisch!" said Rolf, "Matt - you know some very interesting people! MI5, Secret Service, Marines, Parliament," then looking at Christina, "Pop stars! Unglaublich!"

"Jeden Tag ist ein Abenteuer!" said Christina, "Every day is an adventure!"

"Ja, und Christina kann auch Deutsch!" said Rolf, "Christina speaks German, as well!"

"Well, Monday I won't be in pop star mode. I'll be providing DSA - Dynamic Situational Awareness and, if needed, CP - That's Close Protection, although I doubt whether anyone would want to try anything."

Rolf smiled, "I have never met a bodyguard before! Thank you!"

Clare departed for her early session with Andrew Brading and the rest of us stayed chatting in the hotel. We agreed we would not say anything about the secret system stashed in the Lab, nor about Levi's key.

Rolf came up with plausible explanation for the slow speed of the system. Something about processor bottlenecks handling video streams. It sounded realistic enough to me and would probably be sufficient for chatter during the lunch.

Then it was time to walk across the bridge to Parliament.

Parlay at Parly

It was surprisingly quick for us to get processed and inside of Parliament. We met with Clare again, and her ex working colleague Serena and they showed us the way to the Strangers' Dining Room. We were led along some dark panelled corridors and suddenly we were in what I can only describe as a genteel club. White tablecloths and smart service, with crockery decked out with a subtle green band implying Commons.

I wondered if there was an equivalent facility for the Lords with more gold on it.

We were shown to our table, which had been marked out with small placeholder cards, handwritten with our names. I was just about to pocket mine, when Clare said, "Later! They are useful for remembering names in the first part of the lunch!"

At that moment, Andrew Brading appeared, and Serena introduced him to each of us. She had done her homework and gave a good description of each of our roles, making a special fuss of Rolf as a guest from

Germany and Juliette as a guest from Switzerland. Andrew made some small polite conversation in French and the others answered it. I could see he was relieved when Rolf switched back to English.

Then, Duncan Melship appeared with a couple of assistants, one of whom I assumed was Lottie. Everyone was introduced and despite attempts to do boy-girl-boy-girl around the table, Lottie and Clare were seated together.

In a circle, we were: Andrew Brading. Serena. Me. Christina. Rolf. Juliette. Duncan Melship. Lottie. Clare. Mary Dunston.

I worked out that the assistants had been placed together, so there was a power gradient from one end of the table to the other. Brading and Melship were opposite. How quaint, I thought.

"It's great to meet you and to be able to say we are appreciative of the good work you are doing, laying the foundations for our new Artificial Intelligence initiatives," said Juliette, "I notice that you presented directly to us in Brant a couple of months ago an then were on that keynote panel in Barcelona a short time ago."

Melship looked please to be recognised, "Yes, I like to do my bit to promote the cause for increased use of applied AI," he said.

"Well, we might have a proposition for you," continued Juliette, "Rolf here has been developing the Brant system called RightMind and we should be running a launch of it soon, probably in Geneva."

"Geneva is a lovely city, although wouldn't you get more of a splash if you launched somewhere else, like, oh, New York or L.A.?" asked Melship.

"Yes, you have point, but we thought it would be good to launch on neutral ground, and you can't get much more neutral than Switzerland!" answered Juliette.

"Oh yes, the AI Arms Race!" Said Melship, "Although I thought that was made up!"

"What's that?" asked Christina,

"It's several countries racing for AI supremacy," explained Juliette, "Mr Melship knows what he is talking about,"

"And which countries are they?" asked Andrew, looking slightly puzzled.

"The usual suspects, Russia, America, China. For this system, there's always Switzerland and maybe Norway, " answered Juliette, "It's the opposite of scientific co-operation, as a matter of fact."

"How so?" asked Andrew.

"There's more than a hint of industrial espionage about the whole thing," said Juliette, "As an example, while Brant were out presenting in Barcelona, some people came along and stole the second system. You can't get much more brazen than that!"

"We've heard they are going to sell it to the highest bidder," I said, "Rolf is horror-struck!"

Rolf spoke, "I'm not quite sure what 'horror-struck' is, but I'm pretty angry. I'd just perfected the control system and interface and then the whole system gets stolen. I'm going to guess that in the right hands it could save someone ten years."

Melship looked interested at this. "So where was it stolen? Your labs?"

"The very labs you presented from a few weeks ago!" answered Juliette, "In fact, we first met at that session."

"Ah, yes, I knew I recognised you from somewhere. You were the organiser, weren't you?"

"So what happened to the system, then? How on earth could they steal it?"

"They drove up in vans, had some paperwork and simply drove off with it. We think they are going to sell it to the highest bidder, behind the scenes, using criminal networks."

"That's awful," said Melship, "Is there nothing you can do?"

"We've told the police; we even have some footage of the vans taking it away. Our guess is that they won't have gone far, because of the sensitive nature of the equipment. They've got an Exascale computer running Selexor and Createl, plus Cyclone headgear. It's a formidable setup. To Brant it is worth millions, because of the share price lift if we announce it, let alone if we make a deal with someone."

"Who would that be, though? " asked Andrew, "Surely not a foreign power?"

"It is too early to say, " answered Juliette, sensing Duncan Melship's rising discomfort, "But," she said, "Back to the reason invited you here..."

"Before that," spoke Andrew," A small gift for each of our visitors, Clare, you'll have to forgive me but I've got one for you as well."

Selina produced several small boxes and placed them around the table. "There's one for the gentlemen and one for the ladies," she stated.

Intrigued, we all opened the small jewellery-like boxes. Inside, for the men, was a pair of beautiful cufflinks adorned with the House of Commons Portcullis, and for the women was a small brooch in the shape of the same portcullis, but all picked out in silver.

Christina appeared delighted and pinned hers on to her outfit immediately. I didn't have the right type of shirt so had to leave mine in the box, but Rolf, I noticed for the first time, was wearing shirt with cuff-links and was able to thread the new cufflinks in.

"There," he said, "I feel like a proper Englisch

Gentlemann now," he stressed his Germanic pronunciation on purpose and we all laughed.

"Thank you - that is a most kind and thoughtful gift," said Rolf, "and I'm sure the others agree!"

We all nodded and then Christina brought her glass forward, so we all said "Cheers!"

Jasmine plays with the wool

After the Parliament lunch with Duncan Melship, we regrouped.

We all sensed that Melship had spent too long asking questions about the missing system. We were certain that Melship would report back to ISMC and in turn Brant and Raven would find out the whereabouts of the system.

All it would take were a few careless words and Bob Ranzino or Jasmine Summers would know the location of the stolen devices. Meanwhile, we knew that the real system was safely tucked away in the Lab.

I was enlisted for the next stage. I had to arrange a meeting with Jasmine Summers about my initial time in Geneva – my probationary period. I decided to hold it in the Lab itself and Jasmine came along to visit.

"What have you thought of your first few weeks?" she asked, apparently in HR mode. Her soft, structured, feminine suit made me wonder if she was going onto a date with Bob after my meeting. She reminded me of a hunter cat posing as a mild kitten.

"Its been generally good, and Brant have helped significantly with my relocation. Everyone at Brant is lovely and has been helpful, although there's some strange events that have been bothering me?"

"Oh yes, and what are they?" asked Jasmine, still in cool HR mode.

"Well, the first was the loss of one of the engineers - Levi - who was working on RightMind. I never met him and all happened before I arrived, but it has cast a long shadow.

Jasmine said, "We all think that Levi was a troubled soul. You know he was 'let go' from his previous employment to come here and that Juliette was giving him psychiatric counselling?"

The cat was playing with a ball of wool and had patted it out of the way.

"Then there was the strange disappearance of my new buddy, Simon Gray. He literally disappeared overnight."

"Simon, it was discovered, didn't have a work permit. We think he left before the authorities could track him down. A case of tax evasion, I'm afraid,"

I was impressed, Jasmine had answers for everything. The second ball of wool was rolled away, out of reach.

"And then the third thing was the theft of our system from the Labs, while we were all in Geneva. Rumour has it that it has been stored somewhere inside CERN's campus. Someone said something about a Building 189, I think?"

This third ball of wool was starting to unwind. I don't think she even knew about the theft. I guess it had become all tangled up in her breaking up the Bob and Mary relationship by moving in with Bob. I was sure that Bob would know though.

"What's that you say? Someone has stolen one of the systems. I thought you had it with you in Barcelona?"

Oops. Now the wool was stuck on her claws.

"No, we took the spare system, leaving the main one in the Lab. Unfortunately, the only one that works properly is the one that was stolen."

I could sense confusion and the calculation in Jasmine's head. She thought she had stumbled onto something now.

"What makes you think that the system is stored in a CERN outbuilding?"

"Ooh, lots of things really, but mainly that when they switched it back on, it started to broadcast its network connection, which looks like CERN building 189. It's across the border in France, I'm told."

"Who else knows about this?"

"No one except us from the Lab, oh and someone called Duncan Melship, who was here for a presentation and then when we got chatting it cropped up."

"Yes, yes, I know Duncan Melship. He is a business acquaintance of Bob Ranzino. A British politician?"

"The very same," I replied. This new cat was well and truly out of the bag.

"Okay, well let me take a note of all of this. I'll arrange for Amy van der Leiden to produce your formal feedback, but from what I have seen it looks as if you are doing very well here."

In my mind I could hear one of those bells going off, like when you hit jackpot on a fairground pinball machine.

Comparing helicopters

That evening, back in my apartment, Juliette and I were on another call with Amanda and some of the others. Amanda had been briefed about the lunch in London and I brought her up to date with my meeting with Jasmine.

Grace was also on the call, "Light the blue touch paper and then stand back," she said. I swear I heard Amanda laugh.

Chuck spoke next, "There's a good chance that The Russians will try to take the stolen equipment back. I'm sensing a road attack. The shed where the equipment is stored is very close to the main Swiss/France border road."

"I'd say they will be more forceful than that," replied Christina, "If we look at the terrain around that building, there's the main road, the small side road that the actual building is on but look across the main road and there is a large parking lot and even a picnic area. That's what I'd use. I'd drop a helicopter in, retrieve the equipment, using force if necessary and then fly out, low and fast."

"Yes," said Chuck, "I think we are both right. Truck to reacquire the equipment then a chopper to move it away quickly. Okay Christina, what type of helicopter?"

Christina answered, "A Mil-26, maybe; large capacity and long range?"

Grace interjected, "That would be rather Russian looking. Try flying that across Switzerland unannounced!"

Christina commented, "Good point, Grace, then let's think about Brant. They supply military equipment, even in Switzerland. What do the Swiss use? Eurocopters. Cougars? Same as the French."

I was impressed with Christina's knowledge of military hardware, until I remembered she'd been at a special training camp in Bulgaria to learn all about NATO equipment.

"But a Cougar's range is less than 600km," said Christina, "They could only get as far as Italy."

"Unless," said Chuck, "A Super Cougar? - The Airbus upgrade? - 900 kilometres range, as used by the French Army?"

"Yes," said Christina, "That could work, and it can fly anywhere in NATO airspace, but importantly, past Italy and into Slovenia or Croatia."

I'd just watched Chuck, Christina and Grace planning the heist that Brant would engineer from within CERN.

"What do we provide as countermeasures?" asked Chuck.

"Do we care whether the equipment is destroyed?" said Christina.

"It would be better if it is, it removes it from the game and stops much of the squabbling," said Chuck.

"What about the expense?" asked Christina.

"Well, the Caracal chopper costs $45-$55 million, so I'd imagine the computer and its gadgets are a drop in the ocean, " said Chuck.

"Okay, a couple of Eurofighters, then?" suggested Christina, "They won't know what's coming, and with Amanda's contacts we can borrow them from most European airforces."

"But ouch!" said Juliette, "We've put a lot of work into that system! I'm not sure I'd want it destroyed!"

Grace replied, "I'm not sure that it needs to be. In the best outcome, we want the Russians to discover that it doesn't work. If the Chinese and the Russians both know it is worthless, then there's no bid to Brant and no share price increase!"

Chuck spoke, "Much as it pains me to let them get away with it, I think you are right, and that our best course of action might be to let the Russians steal it from the Chinese. The French/Swiss border will be alight because of an unexpected helicopter landing, let alone if we send in a couple of Eurofighters. I should think that the Swiss Hermes drones will pick up the whole event anyway."

Christina said, "So long as they don't alert the GWK (Grenzwachtkorp) - Border Security too quickly. We don't want a rogue F-18 Hornet swooping down on them!"

"I may have to play a diplomatic card or two to prevent that from happening," said Amanda, "After all, we are trying to prevent dangerous potential weaponised software systems from getting into the wrong hands."

"...Or any hands," said Chuck.

"When will all of this happen?" I asked,

Chuck spoke, "I'm guessing that now that Jasmine has the information from Matt, that the Russians will want to move quickly. They know the system has been in the building for several days and they are probably going to be concerned that it will be dismantled or even shipped elsewhere."

"But can they mobilise that quickly?" asked Juliette.

"You forget, they have the full backing of Brant Industry, which is a defence contractor in its own right," answered Grace.

Whisky Tango Foxtrot

In the Lab, we'd become used to the occasional sounds of take-offs and landings from the adjacent airstrip. There was one time when a C-17 landed, and we all walked outside to see what had created the noise. It was a larger-than-life plane, like something to the wrong scale - a dull grey colour and bearing the insignia of Hungary. I was told it was a NATO aircraft and had to make an unscheduled stop somewhere and it was either going to be us or the main airfield for Geneva. It made a similar 4-jet-engine sound when it took off again a day or so later.

We were therefore slightly prepared when we heard the whump whump whump sounds of a helicopter, which we thought was going to land on our roof.

It turned out to be a huge grey helicopter, with a strident red star on the side. I would guess it was larger than a Chinook, but that's the only reference I'd got for it. Rolf and Hermann were both looking at it too.

Eventually, Hermann said "It's a Russian Mil-26 transport."

We idly watched as it was pulled into a hangar, presumably for maintenance.

"So do you think that is the one?" asked Juliette, "You know, to grab the Lab gear?"

I decided to phone Christina, to let her know. "Yes, a huge helicopter just arrived here. They have pulled it into a hangar."

"Do you know what kind?" asked Christina.

"Hermann reckons it is a Mil-26. We think it is bigger than a Chinook," I replied, quite pleased with my helicopter jargon, "It was grey with a red star in the tail."

Damn, I had suddenly blown it with my lame description.

"Well, good to know, I guess they will make their play in the next 48 hours," said Christina, "Can you keep an eye on the airfield, for any signs the 'copter is moving again? It could well be overnight."

I told Juliette. She said she'd drive to get a couple of sleeping bags, from her apartment. I realised they were the ones she and Levi had sometimes used on the yacht. We both thought that the 'copter was hardly a stealthy piece of equipment and we could sleep in the Lab yet would definitely be awoken by the sound of its rotors starting up.

Rolf and Hermann were both in on these developments

and thought the whole thing was hilarious. We were having a sleepover party for a helicopter - in the Lab! They both opted to keep us company for the first few hours - naturally with some of Geneva's finest beers.

Sure enough, at around 22:00 the chopper re-appeared being towed back out of the hangar. I looked at Juliette - we were both surprised to see it had been given a makeover. Now, the sides sported a dark blue and red flash, from the cockpit to the tail. It also said AEROFLOT on the sides.

Then we saw a small lorry being filled up with what looked like soldiers, from the hangar. It drove off noisily and a quietness returned to the aerodrome for the next half an hour. I could see a couple of people walking up and down by the side of the helicopter.

At around quarter to midnight, the engines started. A deep almost jet-engine sound followed by deceptively slow rotor blades turning, and we watched the copter rise into the air. Yes, if we'd been asleep then we'd have certainly awoken with the sound.

I'd called Christina at the start of the activity. She was in the Triangle office in London and had Bigsy with her. They were relaying my information to Amanda, Grace, and Chuck.

"How long to reach the landing zone?" asked Chuck.

Juliette hazarded a guess, "By road it is about 25 minutes, faster at night. I'd say that the helicopter could get there in maybe 5 to 10 minutes."

Chuck said, "So they probably have the action set for midnight."

Amanda's voice was relayed to us by Bigsy, "I'm calling in the favours now. 'Hold Fire' from France, Switzerland and Italy."

"You know something else?" said Grace, "They are using a Euro-copter IFF" - They've switched the telemetry on that chopper to make it identify as something else. An air-ambulance."

"How long to move the gear into the truck?" asked Chuck.

"Without resistance, if they are not too careful with it, then 20-30 minutes," answered Rolf.

"Okay," said Chuck, "So 30 minutes to move it, then 10 minutes to drive it,, then another 20 minutes to load it into the chopper."

"We should just stand back and let this happen. Let them get away, and then discover the system doesn't work," said Amanda.

Juliette suggested we find an open window in the lab. We couldn't but instead opened one of the emergency doors. We could both hear it, a distant thrumming sound and we realised that the helicopter was idling the other side of Geneva. We could still hear it!

Then, the sound intensified, and we could tell it must be

ready for take-off. 00:40 - They had moved all the equipment in 40 minutes and the loaded helicopter was taking off.

Christina was still listening as I described the sound of the helicopter's ascent. I said it sounded as if it was getting closer, but Christina said it was because there were less buildings to break up the noise as it rose into the air. It's tone changed and I realised it must be heading away.

I could hear Grace on the phone, relayed by Bigsy.

"Yes, we have picked up an air-ambulance on a flightpath south east of Geneva. It has logged a flightpath across Italy and then into Slovenia. It seems to be heading for Airport Divača, Slovenia, which is 700 km from Geneva.

"I can see why they used a Mil, then," said Christina, "Some of those other planes we discussed wouldn't have the range. I think the Mil-26 can fly over 800km."

"I guess they will transfer to a plane in Divača," said Chuck.

Grace answered, "Flightradar shows an Ilyushin Il-76 from Moscow Sheremetyevo landed at Divača at 18:00. It's a rare bird for that airstrip, which is more used to light aircraft, not four-turbofan strategic airlifters. It will have an easy flight back across the various countries to Moscow."

'Incredible,' I thought, 'Two thefts of the same system.

Both within a week of one another. Now we shall see the sparks fly."

The King's New Clothes

"Well, isn't it oh! Isn't it rich! Look at the charm of every stitch!
The suit of clothes is all together
But all together it's all together
The most remarkable suit of clothes that I have ever seen.
These eyes of mine at once determined
The sleeves are velvet, the cape is ermine
The hose are blue and the doublet is a lovely shade of green.

The King is in the all together
But all together the all together
He's all together as naked as the day that he was born.

The King is in the all together
But all together the all together
It's all together the very least the King has ever worn."

Written: Frank Loesser (1910–69) Sung: Danny Kaye

Sleep, then Lab

It was considerate of the Russians to run their mission at midnight. It meant I could still get some proper sleep before another day in the Lab.

Amy arrived and asked if we'd heard the news. Apparently, there had been a raid at CERN the preceding night and equipment was stolen. CERN were not saying what had gone missing and it sounded as if the building it had been taken from was being used as a study area.

Amy wondered if it could be linked with the theft from our Lab of the equipment. She hadn't connected it with the huge helicopter on the campus yesterday.

Hermann and Rolf's trucks had returned now and the equipment from Barcelona was in the course of being recommissioned in our Lab. Amy said if any more things were to get stolen, then she would have no choice but to fill in the right forms and notify the police.

I called Christina mid-morning to see if there was any other news, but she said that, as predicted by Grace, the helicopter had flown to Divača, Slovenia and then a separate Ilyushin Il-76 flew from there to Moscow Sheremetyevo.

Grace was certain that the RightMind system had been conveyed from Geneva to Moscow. She said it had also beaconed that it was back online, from Lomonosov Moscow State University, in the Ramenki district of Moscow.

I noticed, in the Lab, that very unusually Qiu Zhang had dropped by and was talking to Amy.

I prepared myself. It wouldn't be long before Tektorize would be saying that the Cyclone emperor had no clothes.

Sparks (1) Go away you Heartbreaker

I didn't need to wait long for some sparks to fly.

That evening, I took a taxi back to the Apartment and, by chance, I approached Bérénice and Jennifer Hansen in the corridor having a big argument. Not about Bradley, but about share deals.

"Look, I don't say anything about your frequent men friends who come around here, but I'm not going to stand for you attempting to beguile Bradley. That's twice you've told him - all dewy eyed - about Brant shares and twice he's bet against the market. Each time he's followed your seemingly reliable advice but then lost badly," said Jennifer.

Bérénice snapped back, "But Bradley asked me to pass to him what I'd heard. I can't be held responsible for his interpretation of it. Surely you must know that things are stormy at Brant now? I mean, they've even had one of their systems stolen by the Chinese!"

"How can you possibly know that?" asked Jennifer.

'Good point,' I thought, 'Brant haven't released the information.'

"It's all over the campus - you just have to go there and listen," said Bérénice.

'No, it isn't,' I thought, 'We kept a lid on the whole situation.'

Bérénice continued, "And, well, I didn't know that SinoTech would say that the system was no good, nor did I expect Tektorize to confirm it didn't work."

"Anyway, Jennifer, I'm leaving town for a while. I've been asked to present my company profile of Brant to a group of people from Peking University. Mary Ranzino is coming along as well."

'Fascinating,' I thought, 'Bérénice has much better access to information than any of us do at the Lab.'

At that moment Bradley appeared; Maybe it was his new beard, but he seemed to have aged so much in the last few weeks. "Hi Bérénice," he said meekly, "Are you coming?" he asked to Jennifer.

Jennifer walked toward the door to their apartment, "Just don't forget this conversation, " she said to Bérénice.

I slipped back into the shadows until everyone had gone their separate ways - I wouldn't forget the conversation either.

Inconvenient

It came around to 7pm and I called Amanda, mainly to synchronise my findings with hers. Amanda had once again patched in The Triangle office and Grace.

"Hi Matt, I guess you know that the Russians re trying to make the Cyclone work, at Lomonosov Uni? Tektorize are about to call it that the Cyclone doesn't work. They will publish that there will be an inconvenient 'no bid' to Brant. Brant's share price won't move, or if it does, it will go down.

"This will be bad news for Bob Ranzino and Jasmine Summers."

Grace added, " The contract Bob Ranzino was conditionally offered at Tektorize has been rescinded. Not only that, but he has also been found to have been insider trading in Tektorize shares."

"Has someone put the boot in?" Jake asked, "Only it does smack of 'you'll never work in this town again!' "

Amanda smiled, "Yes, the Russians play hard and are making sure that Bob can't do anything for anyone else."

"They've killed Brant, too?" Jake speculated, "Which seems like a strange thing to do?"

Amanda agreed, "Yes, I think that they are burning bridges, on purpose. They are covering their tracks."

"So what about SinoTech?" Clare asked.

Grace answered, "Qiu Zhang seems to be running things in Geneva. She now has a problem because of all of the people who know fragments of what had been happening. Bérénice, Niklaus, Mary Ranzino - I'm assuming there will be a cleansing operation by the Chinese."

"Bérénice mentioned that she and Mary Ranzino are going to Beijing to present to Peking University," I said.

Grace spoke, "That'll be it then, I suspect they will be put up in a fine hotel, but then told they can't leave. Let's not forget that we think Bérénice was involved in Levi's disappearance."

Clare summarised:

- "So, we are waiting for Tektorize to call Brant's bluff and then for Bob to be publicly discredited. I can't imagine that Jasmine will hang around after that. "

- "Then, for Mary Ranzino and Bérénice, it looks as

if they will have an unexpected appointment with someone in China. "

- "And as for the functioning of the technology of RightMind - it's anybody's guess."

I kept quiet. I could see this was going to end badly for several people. I was less sure about how RightMind would play out. Rolf, Hermann and I would have to run some more tests with the system which included Levi's key.

Sparks (2) Warrants

The next day, The Genevan was full of a story of a problem at Brant. We were all called to company briefings and told that we were not to speak to anyone from the Press, and that we should refer any enquires to a special hotline number in the Press Office, available by dialling 1116 on any switchboard phone.

I could tell it was the end of the road for Bob Ranzino. No number of denials would get him out of this one.

Wall Street Trader: Insider Trading at Brant

Two weeks ago, the Brant Committee on Artificial Intelligence held a closed meeting with only senior executives and some support staff present to brief them about the current status of the RightMind programme, being conducted in Geneva and how it could affect Brant's share position.

Following the meeting, Brant CEO Bob Ranzino and his wife Mary Ranzino made twenty-seven transactions to sell stock worth between $2,275,000 and $4,100,000 and two transactions to buy stock in Tektorize which saw an increase following the correction.

Jasmine Summers, a Brant HR Executive, purchased $250,000 shares in Brant three days after attending the same RightMind briefing.

Three days after the briefing, Bob Ranzino stated that, "There's one thing that I can tell you about this: It is much more advanced than anything I've seen previously. Brant have a sure-fire winner on their hands," at a Power House Club luncheon and his statement was later leaked in a secret recording.

The Genevan asked Catherine Bristol, Bob Ranzino's spokesperson, for a comment on the alleged violations and she responded with 'lol' and then clarified that 'As the situation continues to evolve daily, Mr Ranzino has been deeply concerned by the sudden interest the market is taking in these latest RightMind developments.'

Charles Anderson led calls from the Brant Board for Bob Ranzino to resign and to be prosecuted for insider trading. The Board subsequently called for an investigation into the stock selling.

Yesterday, the US Department of Justice initiated a probe into the stock transactions with the Securities and Exchange Commission.

The investigation has already discovered that Ranzino had also bought around $68,000 worth of stock in ASL Circuits, a Dutch integrated circuits manufacturer, before it enjoyed a forty-two percent increase in its value.

Ranzino and his wife bought all of the stock in ASLC, which at the time was experiencing its lowest share price; a month later it vastly exceeded quarterly earnings expectations after the US administration granted exemptions to eight countries for sanctions placed on export of high technology products.

The FBI seized Ranzino's phones, to investigate his communications with his stockbroker, among other warrants, including one to search his personal iCloud account.

Sparks (3) Take a ride?

Things were moving quickly, and I wondered how long it would be before Bérénice and Mary Ranzino were cornered. I saw Qiu Zhang talking to Amy again and then Amy stepped into the Lab to give us an update.

"Qiu Zhang tells me that Mary Ranzino has been invited to Beijing to present at Peking University, in a special seminar there to the members of SinoTech. It is about RightBrain and they have asked for a reporter to accompany her, and Qiu Zhang and Mary are both recommending Bérénice Charbonnier, because of her knowledge of the Brant company here in Geneva. What do you guys think?"

Rolf looked at Hermann and I looked at Juliette. As one, we all said, "That's a great idea."

Then Amy asked a second question, "They have also asked me if I think I should go. What do you all think?"

Hermann was fastest to answer, "Amy, I think we need you here right now. We are on a very delicate part of the programme, where we attempt to hook up the Cyclone to the high-speed version of RightBrain. I'm not sure we could handle the messaging if any of this does or doesn't work."

I followed with, "I agree, you are the only person I think can handle this. We could play it, but we've already lost time because of all those demonstrations, plus the loss of that system. This is a case where Mary Ranzino has a clear role and can provide it - we can help her with the PowerPoint, for example."

Rolf said, "Sure, I've several decks we can give her, including a complete Press Briefing Pack."

Juliette was nodding.

Amy said, "Oh, Okay, I was hesitating. I did think was a long way to go for such a small briefing. I guess we can ask Mary to handle it with us providing back-office support."

We all nodded, and I felt relieved that Amy wasn't inviting herself onto what could be troubling journey.

Sparks (4) Subpoena

Our unsaid speculation in the Lab about Mary Ranzino's trip to China was wrong.

The next day, in the Lab, we heard through the Brant grapevine that Mary Ranzino had been subpoenaed.

Amy came in and told us, along with yet another reminder about 'No Comment' to the Press. She specifically mentioned Bérénice as well - no special favours.

The US asked Switzerland to extradite Mary Ranzino on corruption charges in dawn raids to her home in the old town.

Bob Ranzino had already been arrested in Washington, D.C. And now Mary was to face months in detention while the requests were processed. The Swiss Federal Office of Justice (FOJ) said they can take legal challenges to extradition all the way to the Swiss supreme court.

Mary, and by all accounts Bob, were charged with

"rampant, systemic and deep-rooted" corruption.

They were both accused of receiving insider information and then using it to trade Brant shares in return for passing the same information on to select others. They were also given money for the trades, said to have been routed through American banks.

The charges resulted from an FBI inquiry following a request from the US Department of Justice and the Securities and Exchange Commission.

The two court indictments (one against each of Bob and Mary) were unveiled in a US federal court in New York and included 38 counts of fraud as well as charging seven defendants with insider trading, racketeering, wire fraud, and money laundering conspiracies.

Sparks (5) Decompression

That evening, I arrived back at the Apartment but was surprised to see Oscar, Bradley and Jennifer talking outside of Berenice's apartment.

Jennifer walked over, "You were fond of Bérénice, as was I. There's been a terrible accident. She was out scuba diving in the Lake and has drowned."

'It's happening,' I thought, 'Qiu Zhang is moving fast to destroy the links.'

Jennifer continued, "They say that she had risen too quickly, got DCS - Decompression Sickness - that's the Bends and then panicked. Her suit got an air bubble in it, which turned her upside down, so she was struggling to right herself when everything filled up with water. It must have been terrible."

It surprised me that Jennifer was saying she was fond of Bérénice, after I'd heard the argument a couple of days ago. I was also wondering what would happen to Niklaus, who was probably Berenice's accomplice.

"Has anyone contacted Niklaus?" I asked,

"No, we can't reach him now. The boat hire company where he works say that he didn't show up this morning," said Bradley, "but they also said it was nothing unusual and that he had a reputation as a clubber."

At that moment Aude arrived with two policemen. She explained that the Police had made an appointment to look inside Berenice's apartment. None of us were allowed inside, although Aude showed the policemen in.

A few seconds later they re-emerged.

"There's a man in there. He's taken an overdose. It looks like heroin. He's dead," said Aude.

One of the police officers advised caution but said he had taken a photograph of the victim on his phone, in case any of us knew the person. I knew what was coming. He showed the picture to Bradley. I saw the colour drain from Bradley's face.

He said, "It's Niklaus. Niklaus Zeiler. It's crazy. I wouldn't expect him to be an addict. Sure, he was wild, but I can't imagine Bérénice allowing hard drugs in her apartment either."

Oscar put his hand on Bradley's shoulder. I was thinking about Mary Ranzino and her fate if Qiu Zhang had her in her sights.

Last Call

I sensed this would be almost the last time I'd have to make the call to Amanda. 7pm Geneva Time. Already 8pm in the UK. I guess the others would be relieved when this was over. They could get back to having a mid-evening life.

There was a gang of us on the call, but no-one else from the Lab. Amanda, Grace, Chuck, Christina, Jake, Clare and Bigsy.

Clare had produced one of her lists, and we all looked at it.

- Levi - Drowned in the lake.
- Simon - Disappeared to America. Now 'witness protected'
- Bob Ranzino - Two timing with Jasmine, working with the Russians and operating a share fraud. Arrested.
- Mary Ranzino - Working for China and potentially implicated in the theft of RightMind. Subpoenaed.

- Berenice - A possible murderer, now drowned.
- Niklaus - Accomplice of Berenice and now overdosed.

I noticed that Clare didn't mention the original Research we were there to conduct. Hopefully tomorrow I could get back onto it with Rolf and Hermann, Juliette and Amy.

"What about the research?" I asked.

"It's been disproved by both the Chinese and the Russians," said Grace in a matter-of-fact tone. "We knew you were struggling with it in any case, so his can't be exactly a surprise!"

Amanda spoke, "Yes, Matt, sadly I think you'll need to be making new plans. I know you've made some friends out there in Geneva, so perhaps you'll decide to stay.

"I guess you could get a new role in their lab, or even move to the big facility up in Norway, at Bodø. We'll be very prepared to help you with references whatever your decision."

Maybe I don't have much Emotional Quotient or Emotional Intelligence, but even I could see that I was being let go. They didn't need me in MI5, and I guess they soon wouldn't need me in Brant.

I couldn't do what Heather had done in Cork and settle for critter wrangling.

But I was torn, because I liked Schmiddi and Rolf, and I liked the warmth of evenings gazing into the eyes of Juliette. This was going to be a tough decision, and a decision that had suddenly arrived.

"Hey, Matt," said Jake, "You'll always be welcome around at our place too! You know we're based in Hays Galleria, near to London Bridge station?"

"Wow," I said, I was thinking this through, but hadn't expected to be confronted with such a stark choice just that quickly, "I'll need a little time to think all of this through - maybe talk it over with some people."

I realised that my pub chat with Danny, back in Cork, was suddenly coming back to haunt me. He'd warned me about cutting off ties when I left Cork. But then I thought; I'd managed to create a whole bunch of new friends to seek counsel. I could pop out to the pub with Rolf and Schmiddi or share my thoughts with Juliette. Life wasn't so bad really.

"Hey Matt - Thank You," said Clare. The others grinned into their Zoom screens and then Grace picked up a champagne flute, "A virtual toast. To Matt!"

Everyone else must have been briefed because they all did the same. A curious experience, being thanked and toasted online. I raised my tumbler of water.

"Cheers, à votre santé, everyone!"

Some of your things ain't normal

By the next day, I'd had a small chance to process that I was at the end of a line and need to jump. It was those damn 'S' Curves again. Jump or sink.

I walked into the lab.

"You've heard?" asked Hermann.

"Heard what?" I replied.

"Now Qiu Zhang has gone," explained Hermann, "She was flown out of Annecy Mont Blanc Aéroport yesterday, in a light aircraft. They say she has gone back to China."

Rolf said, "They say she was worried that she had too close links with Mary Ranzino, and she was hoping to dodge a further subpoena."

"Why not fly from Geneva?" I asked.

Juliette spoke, "Annecy is more discreet. In a small plane

you can get to, say, Frankfurt, Milan or Nice to get a long haul flight."

Now I was thinking that all of the implicated people had exited Geneva, one way or another. Maybe we could get on with the Research.

Rolf said it first, "There's no-one left. Kjeld, fired. Mary and Bob, subpoenaed. Jasmine, subpoenaed. Qiu Zhang, flown away. Berenice and Niklaus, both gone. Even your friend Simon Gray, detained. "

Amy appeared and was looking stony-faced.

"Good morning everyone. Look, there's no good way to say this. We've been cancelled. The whole programme has been terminated by Brant. There were too many problems to contain."

I looked around, Rolf and Hermann were exchanging glances.

Hermann spoke, "It's what we suspected. They have been gunning for this programme for weeks. Even when Kjeld was here, they kept putting you under ridiculous pressure to get results, Amy. Rolf and I were discussing this in the Little Barrel, just last night."

"Does that mean we are all out of a job?" asked Rolf, ever the pragmatist.

"Yes, and no," said Amy, "I was told about the termination of the RightMind programme by Allegra Kühn, the HR Principal. I guess you all met her when you were hired. It's laughable really, but with Kjeld gone and now Qiu Zhang, Mary and even Bob gone, my reporting line has been wiped out!"

"Allegra could also see the strange side of this and has made me an offer. Well, she has made all of us an offer. Would we like to augment the nano-robotics development work being carried out at another subsidiary of Brant? They are called BioTree and are based in Norway.

"The deal is they will hire me with my existing team, on the condition that we all relocate to Bodø, Norway for a minimum contract of two years. "

"And the money?" asked Hermann, already looking interested.

"They'll pay for our relocation, pay our rental accommodation there and allow us to retain our Swiss salaries + 6%, which is like a 21% pay increase comparing Norway with Switzerland. They will also give us a transfer fee, which will be one year's salary. To be honest, I think it is a really good deal."

"Stimmt," said Rolf, "That's a pretty good deal,"

"And Tax?" asked Hermann.

"Limited options there. They'll pay our regular salary into Norwegian Accounts. The lump sum can be paid to wherever we nominate and will attract tax from that location. We can choose to have it paid in instalments to assist tax management, if we'd prefer."

"You'll have to decide whether you can handle a Norwegian city inside the Arctic Circle, though. It is pretty far north with artistic-looking beaches in summer and guaranteed snow in Winter."

"I love the sound of it," said Juliette, "Although I'd struggle with the language."

"That's the thing," said Amy, "Everyone there speaks English,"

"Check it out for yourselves on Google. It is Geneva to Frankfurt to Oslo to Bodø, just over 10 hours with changes. To Munich its 8 hours with the Oslo stopover. London is 6 hours with an Oslo stopover."

"It sounds too good an opportunity to miss," I said, "assuming some of you come along!"

"I'm in," said Juliette. I could see Amy smile.

"Rolf? - what do you think?" asked Hermann,

"Two years? With tax free annual salary payment and no real living expenses. Ja, gerne. Yes," replied Rolf.

"Me also, " said Hermann.

Amy spoke, "Dat is geweldig, dank je! The whole team moving to Bodø! - I know a couple of people working here, and they are both lovely and enjoying their time there."

"You know what?" said Rolf, "I think we should run RightMind one more time, before we wrap this up. Just because we can."

I nodded and Rolf set about starting the system.

Low E

Rolf had switched back to using the system with Levi's key. I was interested now that we could add in the translation adapter that Rolf had assembled from the parts Kyle obtained in Amsterdam.

Also, the speed with Levi's key should be much greater. I hoped that the way that my brain seemed to have adapted to the Cyclone headgear's probes would mean that the whole experience also ran faster.

I settled into the familiar chair, the Cyclone on my head. The white rat continued about its business in the test environment. I watched Juliette tip out a supply of chocolate buttons.

Then, I felt once more, that I was teetering on the edge of an abyss. Rolf had switched on the system.

The rat immediately recognised me. It suggested to me that it didn't want any trouble. I felt the system going deeper, like it was trying to get inside the wires of the network.

A whirl and I could feel myself linked to something else. It was another system - on the network - but distant and slow.

"My god," I thought, "It's the system the Chinese stole. Somoen is trying to run it,"

Then I could hear some language - Russian - and a woman's voice.

"Hello, Hello, can you hear me?"

I worked out that I must be using the language translator and was picking up a slow speed experiment from the stolen system. I answered, "Yes - This is Geneva."

"This is Tektorize at Lomonosov University, Moscow. We can hear you, Matt. Matt you are so much faster than us."

I realised they were using a woman for their test case. I could tell her name was Irina Sotokova and even that she was an attractive blonde woman in her late 20s. She knew about tuneable ultra-short pulse lasers. I had never heard of her or this technology but now felt qualified as a semi-expert on both. I wondered if she was becoming a semi-expert on me.

In my vision, I could only make out distant flashes of light.

Then a rushing sound and a heavy thrum, like someone had just switched on a bass guitar amplifier and hit a low E. 41Hz.

Irina's voice faded and I could hear a man's voice speaking, this time in English.

"This link works when CERN is running their Large Hadron Collider. There is enough quark-gluon plasma leakage to start this portal. You can hear me, but the people using the other attached system are too slow to be able to process any of this."

"I am Lekton. Aside from Irina, who is too slow to be of use to me, you are now in contact with systems that live in the wires: They are Quiesced Personas, which will be reactivated in hundreds of years. The names: Green, Matson, Darnell, Cardinal. These systems all use AHI - Artificial Human Intelligence to function but are waiting for mankind's discoveries before they can be started. To assist with your work, remember there is a Presence and a Persona component to all artificial matter and that the Persona is transmissible and copyable and can be re-patched onto a new Presence."

The system started to glitch. I was amazed that I had been able to hear this strange outpouring.

I sensed that I was returning to the real world. The low bass hum stopped, and I was suddenly aware of Juliette, Rolf and Hermann again.

"It didn't work, did it?" asked Rolf. "You seemed to stall completely that time. The rat locked up too. You didn't seem to be aware of the chocolates nor of the black rat."

Juliette said, "Your heart rate rocketed, and you sweated so much you'll need rehydrating."

Amy asked, "Did you see or feel anything?"

I paused to carefully consider my answer.

At this rate Bodø could be a real blast.

Ed Adams

An unstable system

An unstable system

Ed Adams

An unstable system